Veterinary Guide
to Horse Breeding

Kjersten Darling, D.V.M.
and
James M. Giffin, M.D.

HOWELL
BOOK
HOUSE

This book is printed on acid-free paper.

Howell Book House
Published by Wiley Publishing, Inc., Hoboken, New Jersey
Published simultaneously in Canada

Limit of Liability/Disclaimer of Warranty: While the publisher and the author have used their best efforts in preparing this book, they make no representations or warranties with respect to the accuracy or completeness of the contents of this book and specifically disclaim any implied warranties of merchantability or fitness for a particular purpose. No warranty may be created or extended by sales representatives or written sales materials. The advice and strategies contained herein may not be suitable for your situation. You should consult with a professional where appropriate. Neither the publisher nor the author shall be liable for any loss of profit or any other commercial damages, including but not limited to special, incidental, consequential, or other damages.

For general information about our other products and services, please contact our Customer Care Department within the United States at (800) 762-2974, outside the United States at (317) 572-3993 or fax (317) 572-4002.

Wiley also publishes its books in a variety of electronic formats. Some content that appears in print may not be available in electronic books. For more information about Wiley products, visit our web site at www.wiley.com.

Library of Congress Cataloging-in-Publication Data

Darling, Kjersten
        Veterinary guide to horse breeding / Kjersten Darling and James M. Giffin
                p.cm.
        Includes index.
        ISBN 0-7645-7128-1 (paper: alk. paper)
        1. Horses—Breeding. 2. Horses—Reproduction. 3. Horses—Generative organs—
Diseaases. 4. Veterinary obstetrics.
I. Giffin, James M. II. Title.
SF291.D27   1998
636.1'082—dc21

                                                          98-27929
                                                          CIP

Manufactured in the United States of America

10 9 8 7 6 5 4 3 2

CONTENTS

Chapter 6—ARTIFICIAL INSEMINATION—87

Chapter 7—PREGNANCY—103

Chapter 8—MARE INFERTILITY—115

Chapter 9—ABORTION—141

Chapter 10—STALLION INFERTILITY—155

Chapter 15—CARE OF THE NEWBORN FOAL—225

APPENDIX—255

GLOSSARY—263

INDEX—270

INTRODUCTION

The *Veterinary Guide to Horse Breeding* was written because of an expressed interest by horse people who would like to learn more about this subject. Many horse owners and breeders simply do not have access to a source of information that is both comprehensive and easy to read. This is particularly true for those who are just getting into breeding, even though they may be highly accomplished in other aspects of horse management.

At the same time, progress in veterinary medicine and especially in the field of horse breeding makes it almost a necessity for even experienced horse breeders to have an up-to-date source of information to consult in order to keep abreast of developments.

In writing this book, we have attempted to provide information on all aspects of horse breeding, including choosing the horses; when to breed; breeding methods; the diagnosis and management of pregnancy; foaling, and complicated foaling; and what to do while awaiting your veterinarian. Additional chapters cover subjects of infertility in the mare and stallion, birth control, artificial insemination, and assisted reproductive techniques including embryo transfer. There is also a large chapter devoted to care of the newborn foal and treatment of neonatal diseases.

The expanded role of cooled-transported semen has changed many of the fundamental aspects of horse breeding and created new challenges and opportunities. The entire subject, including how to test the stallion's semen for cool storage, how to order and ship semen, and when to inseminate the mare, is extensively covered and explained in set-by-step detail.

To clarify medical and technical terms that may be encountered by the reader, we have included a glossary in which such words are defined in simple language. Many of these terms will be found in italics.

This book is not intended to be a substitute for professional care. Book advice can never be as helpful or as safe as actual medical assistance. No text can replace the interview and physical examination, during which the veterinarian elicits the sort of information that leads to a speedy and accurate diagnosis. But the knowledge provided in this book will enable you to work with better understanding and in a more effective partnership with your veterinarian.

The combined efforts of many people have made this book possible. We are again indebted to Susan Stamilio (SKS Designs) for the many fine anatomical drawings throughout this book.

A special note of thanks is due to Dick and Willa Sell of Real Impressive Paints (Montrose, Colorado), to Penny Walsh of Walsh Quarter Horses (Montrose, Colorado), to Billy Scales and Shelby Mighell of B & S Quarter Horses (Ridgway, Colorado), J Bar D Studios (Grand Junction, Colorado), and to the many gracious people who contributed helpful advice and allowed us to photograph their horses.

We also acknowledge with appreciation the numerous researchers, clinicians, and educators whose works have served as a source of information for this book. Among them are *Equine Reproduction*, by Angus O. Mckinnon, BVs, MSc, and James L. Voss, DVM, MS, 1993 (Lea and Febiger); *Current Therapy in Equine Medicine 3*, 1992 (W.B. Saunders Company); *The Horse*, by Calvin Kobluk, DVM, Trevor Ames, DVM, and Raymond Goer, DV Sc, 1995 (W.B. Saunders Company); *Equine Clinical Nutrition: Feeding and Care*, by Lon D. Lewis, DVM, Ph.D., 1995 (Williams & Wilkins); *Nutrient Requirements of Horses*, Fifth Edition, 1989 (National Research Council); *Student's Guide to Equine Clinics*, by Chris Pasquini, DVM, and Susan Pasquini, 1993 (Sudz Publishing); *The Stallion*, by James P. McCall, Ph.D., 1995 (Howell Book House); *Blessed Are the Brood Mares*, Second Edition, by M. Phyllis Lose, VMD (Howell Book House); *The Complete Book of Foaling*, by Karen E.N. Hayes, DVM, MS, 1993 (Howell Book House); and the *Horse Owner's Veterinary Handbook*, Second Edition, by James M. Giffin, M.D., and Tom Gore, D.V.M., 1997 (Howell Book House).

We would also like to express our thanks to the publishers of *Equine Reproduction*, 1993 (Lea and Febiger), for giving us permission to use several photographs from that work.

To Sean Frawley and Madelyn Larsen at Howell Book House, we once again send a special thanks.

AUTHORS

Kjersten Darling, D.V.M.

Dr. Darling, a specialist in equine reproduction, graduated from U.C. Davis Veterinary School in 1985 and received her postgraduate education at the University of Pennsylvania School of Veterinary Medicine. After completing her training, Dr. Darling took a two-year position with the Alamo Pintado Equine Medical Center in Santa Barbara, California, a large veterinary establishment that serviced over 300 mares yearly. Later she served as resident veterinarian at the Westerly Training Center, a Thoroughbred breeding and training farm specializing in sports medicine.

Dr. Darling opened her own practice in 1989 and currently is the chief of veterinary services at Golden Eagle Farms in San Diego, California. She is the mother of two boys, and in her spare time raises and shows paint horses.

James M. Giffin, M.D.

Dr. Jim Giffin received his medical degree from Yale University School of Medicine and completed his surgical residency at Barnes Hospital in St. Louis. Dr. Giffin has maintained a longstanding interest in dogs, cats and horses. He is the co-author of several award-winning books published by Howell Book House, including *The Complete Great Pyrenees*, *The Dog Owner's Veterinary Handbook*, *The Cat Owner's Veterinary Handbook*, and *The Horse Owner's Veterinary Handbook*. Dr. Giffin retired from the practice of surgery in 1995 to devote his time exclusively to writing. He and his wife, Diane, and three-year-old daughter, Sarah, live in the mountains of western Colorado.

THE BREEDING MARE

SELECTING THE BROODMARE

The question is often asked, "Is my mare good enough to breed?"

The answer will depend, in part, on what you are trying to accomplish. The minimum objective of any breeding program is to preserve the qualities of the breed type and produce foals that combine the best qualities of sire and dam. If these basic objectives are not met, the commercial value of foals and the reputation of the breeding program will suffer.

The purposes for which the breeding is intended will greatly influence how much importance you will want to place on the mare's bloodlines, her performance record, appearance, temperament, conformation and soundness.

BLOODLINES

Pedigrees are important because they indicate what the breed is naturally inclined to do. Successful bloodlines produce dominant traits that have proven to be worthy through repeated testing in the field. It is important to learn as much as you can about the bloodlines of your mare, as well as those of the stallion you may have selected. Doubling up on proven bloodlines is a successful strategy that helps to fix desirable traits and eliminate weak ones.

A self-confident mare with a sweet disposition is a pleasure to own.

PERFORMANCE RECORDS

Performance titles and championships do indicate merit and are especially important when the intended purpose is to produce winners at the racetrack or in other competitive events. A filly with a record of accomplishment is a good bet to produce a promising foal. Outstanding performance records, however, do not always guarantee the overall superiority of the offspring since numerous variables can influence results. Similarly, an outstanding mare may have no performance record simply because she did not compete.

APPEARANCE

Correct type is important for beauty and appearance. *Type* is a set of distinguishing physical characteristics that gives the purebred horse the defined look of its breed. The correct breed type is described in the individual breed standards. Your broodmare should be a very good or excellent example of her breed.

Conformation refers to the way in which the various angles and parts of the horse's body agree and harmonize with each other. Standards for registered horses describe the ideal conformation for each breed. These standards are based in part on aesthetic considerations, but take into account primarily the purposes for which the breed is intended. Horses with good conformation are sounder than horses with poor conformation—and are less likely to suffer from lameness. Good conformation is tremendously important in broodmares, since conformation tends to be passed on to offspring.

Experience is required in judging type, conformation and soundness in horses. If you do not have the expertise to objectively evaluate the qualities of your prospective broodmare, it is a good idea to obtain the advice of a knowledgeable breeder or equine veterinarian.

TEMPERAMENT

A broodmare should have a good disposition. Many horse breeders feel that the mare contributes more to the temperament of the foal than does the stallion. Although some temperament faults, such as a mean spirit, can be caused by abuse or incompetent handling, others are genetically determined. Moreover, the mare with an aggressive or excessively shy disposition is more difficult to breed and often does not make a good mother.

In summary, poor type, poor conformation, unsoundness and ill disposition are hereditary traits. Don't make the mistake of thinking that a mare with any of these qualities can be bred to a superior stallion and produce outstanding foals. The usual outcome of such a breeding is a foal of indifferent quality that adds little to the program. Remember that the costs of feeding, housing, maintaining, treating and breeding your mare will greatly exceed her initial cost in all but the most unusual circumstances. It therefore makes good economic sense to start with the best broodmare you can afford.

MARE ANATOMY

The mare's reproductive tract is composed of the ovaries, uterine tubes (oviducts), uterus, vagina and vulva. The pineal gland, hypothalamus and pituitary gland control the hormonal phases of the reproductive cycle.

The ovaries are bean-shaped and vary in length from 1.5 inches during the nonbreeding season to 3 inches during the period of sexual activity. In addition to producing the eggs, the ovaries produce the sex hormones that prepare the reproductive tract for breeding, fertilization and the support of pregnancy.

The outer surface of the ovary is covered by a thick capsule that prevents growing follicles from expanding outward. As a follicle within the ovary matures and prepares to ovulate, it is pushed toward a chute-like recess in the lower pole of the ovary called the ovulation fossa. The end of the oviduct covers and encloses the ovulation fossa. Ovulation thus occurs directly into the oviduct.

The uterus is composed of a cervix, body and two uterine horns. The body of the nonpregnant uterus is about 10 inches long. The cervix is the muscular outlet of the uterus that forms a channel between the uterus and vagina. It is about 3

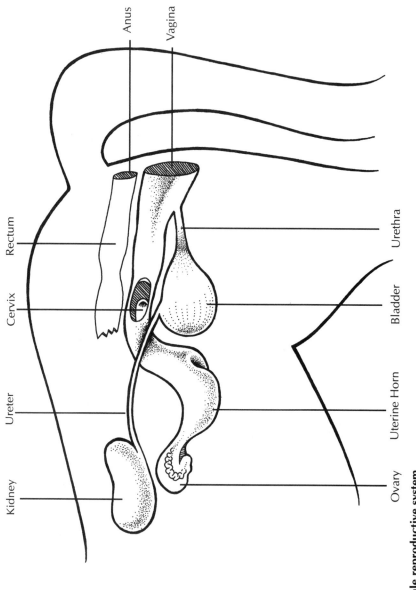

Female reproductive system.

inches long. The uterine cavity is lined by a layer of glandular tissue called the endometrium. The purpose of the endometrium is to nourish and support the growing embryo.

The vagina is a tubular structure about 18 inches long as measured from the vulva to the cervix. The outer one-third of the vagina is called the vestibule. The opening of the bladder (urethral orifice) is located close to the junction of the vestibule and vaginal canal. The vulvovaginal fold is a flap of wrinkled tissue that runs horizontally across the vaginal floor obscuring the urethral orifice. The fold is reinforced by vaginal constrictor muscles. The vulvovaginal sphincter mechanism is a major defense against bacterial contamination of the uterus. It also directs urine toward the vaginal opening during urination. The vaginal canal ends in the vaginal fornix.

The vulva consists of the two labia (lips) of the vagina and the clitoris. The body of the clitoris contains erectile tissue. Beneath and on either side of the clitoris is a depression called the clitoral fossa.

Just above the body of the clitoris is the median clitoral sinus, a pocket about 1 cm deep. On either side of the median sinus are two shallow sinuses. These lateral pockets are present in most but not all mares. All three sinuses contain smegma and may harbor bacteria that can be transmitted to the stallion during breeding.

BREEDING SOUNDNESS EXAMINATION OF THE MARE

The principal purpose of the breeding soundness examination (BSE) is to identify and correct any problems that may interfere with breeding, conception and the maintenance of pregnancy. The exam should be done annually, well before the breeding season. This allows time to treat uterine infections and correct other problems that may exist. This greatly improves the mare's potential to become pregnant early in the breeding season.

Before purchasing a mare, it is a good policy to do a BSE to make sure there are no reproductive problems. A BSE is also performed as part of an infertility workup.

Routine procedures performed during a BSE include rectal palpation, vaginal speculum examination and occasionally *ultrasonography*. Endometrial biopsy and fiber-optic examination of the uterus are indicated for infertility evaluation. Chromosome analysis may be of value for unexplained infertility in maiden mares.

HISTORY

Data on the mare's reproductive status (whether currently pregnant, lactating, barren or maiden), her age, cycling pattern and past breeding successes or failures provides significant information about her present breeding potential.

Maiden mares are mares of any age who have never been bred. Young maiden mares can exhibit abnormal estrous cycles and may express fear of the stallion. A barren mare is a mare who has been pregnant at least once but is not pregnant now. Older barren mares who fail to conceive after two or more breedings usually have an infertility problem related to a chronic uterine infection.

Mares over 15 years of age are often less fertile because they are more likely to have acquired multiple uterine infections associated with breeding and foaling. If the health of the uterus is good, however, age alone does not lower the fertility rate.

A mare with a record of consistent breeding success is unlikely to have a reproductive problem identified on routine BSE. However, the mare with a history of recurrent pregnancy losses or inability to become pregnant is a problem and will need a more complete veterinary examination (see chapter 8, "Mare Infertility").

GENERAL HEALTH AND APPEARANCE

The general health of the mare is important. A mare suffering from a chronic disease, such as heaves or congestive heart failure, is a poor candidate for breeding. A crippling disease such as severe *laminitis* may become so painful during late pregnancy that carrying a foal becomes impossible. Less-critical tendon and joint injuries often improve with rest and anti-inflammatory drugs. If possible, breeding should be postponed until the condition heals.

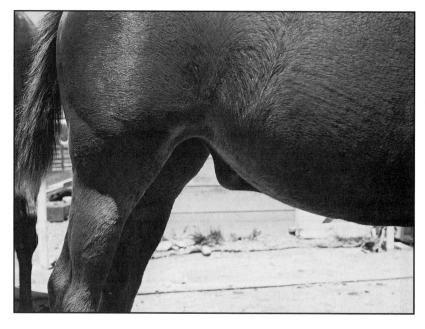

An umbilical hernia should be repaired before breeding.

Umbilical hernias are common in fillies and colts. Most hernias close spontaneously or are repaired before one year of age. An umbilical hernia in an adult mare is likely to enlarge with pregnancy and trap a loop of intestine. The hernia should be repaired before the mare is bred.

A dental examination should be part of the routine BSE. It not only identifies problems that need to be corrected before breeding, it also helps to verify the age of the mare.

Extremely thin and extremely obese mares may have a more difficult time conceiving. These individuals should be placed on a diet with exercise to achieve a better body condition score (at least moderately thin or moderately obese). For information on body condition score, see the appendix.

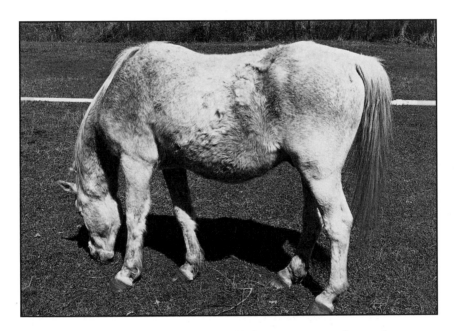

A mare who retains her winter coat probably has not yet started to cycle.

Some maiden and barren mares exhibit an abnormally long winter coat (hirsutism) as the breeding season approaches. This winter coat will be lost before the mare begins to cycle. Thus a hairy coat that continues into spring suggests that a mare has not yet begun to cycle regularly. Rarely, hirsutism is caused by a pituitary tumor.

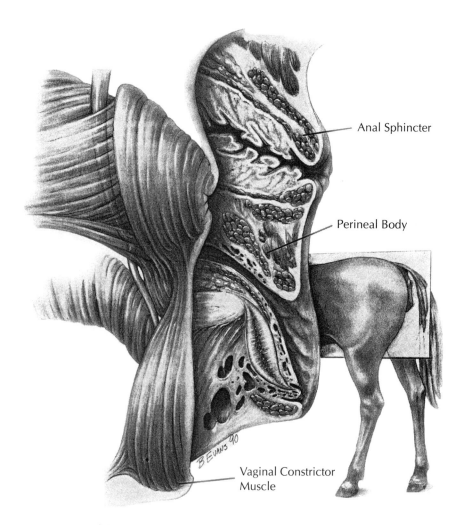

Anal Sphincter

Perineal Body

Vaginal Constrictor
Muscle

Cross-section of the mare's perineum. Angus O. McKinnon, BVSc, and James L. Voss, DVM, *Equine Reproduction*, eds. Lea & Febiger (1993). Reproduced by permission.

PERINEAL EXAMINATION

The type of *perineum* the mare posseses has a direct bearing on her fertility. A mare with poor perineal conformation and a *windsucking* or *pneumovagina* is highly predisposed to vaginal and uterine infections. Such a mare often does not become pregnant, or if she does, she aborts early in pregnancy due to scarring of the uterine cavity. The recognition of pneumovagina and its treatment are discussed in chapter 8.

The clitoris is examined for signs of enlargement. An extremely enlarged clitoris resembles a short penis. Clitoral enlargement indicates recent testosterone administration, an androgen-producing tumor of the ovary, or a sex chromosome abnormality. The most common cause of the latter is the male pseudohermaphrodite, possessing testes and a female *karyotype*.

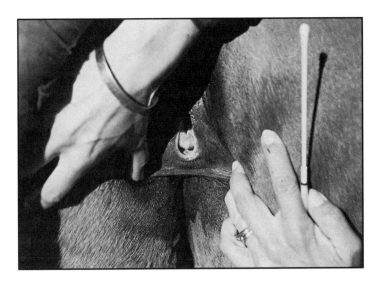

Location of the clitoral sinuses and fossa.

CLITORAL/VESTIBULAR CULTURES

Your veterinarian will take cultures from the vaginal vestibule, clitoral fossa and clitoral sinuses as part of the routine BSE to screen for bacteria that can cause veneral infections. The bacteria of primary concern are *Klebsiella pneumoniae*, *Pseudomonas aeroginosa* and *Taylor equigenitalis*. These organisms can be passed to a stallion from an asymptomatic mare during breeding.

RECTAL PALPATION

Rectal palpation is informative as to the size, shape and location of the ovaries, and the size and consistency of the uterus and cervix. The hand and arm in a well-lubricated shoulder-length glove is inserted through the anus into the rectum. The pelvic structures are examined through the wall of the rectum with the fingers and cupped hand, much like feeling a tennis ball covered by a towel. The mare should be restrained in a palpation stock to prevent sudden movement and injury to the examiner. A nose twitch and/or a tranquilizer can be considered for nervous mares and known kickers.

Injury to the wall of the rectum or colon during palpation is rare but does occur. When it does occur it is often fatal. Accordingly, rectal palpation should be performed by experienced personnel.

The uterus is examined for size, position and degree of firmness. A large soft uterus suggests uterine infection.

The ovaries are examined between the fingers and thumb. The size of the ovaries, along with the presence or absence of ovarian follicles, should correlate with the stage of the estrous cycle. Small ovaries are hormonally inactive and indicative of *anestrus*. A mass on or near the ovary may be a paraovarian cyst, hemorrhagic follicle or ovarian tumor.

The cervix is palpated by compressing it against the floor of the pelvis with the tips of the fingers, feeling for lacerations and scarring.

VAGINAL EXAMINATION

This is an important part of the BSE. The mare should be restrained as described for rectal palpation.

Vaginal examination carries a risk of introducing air and bacteria into the uterus. Accordingly, all vaginal exams should be performed using sterile technique.

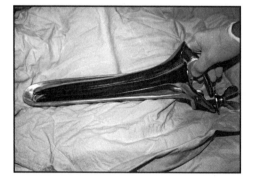

The bivalve vaginal speculum.

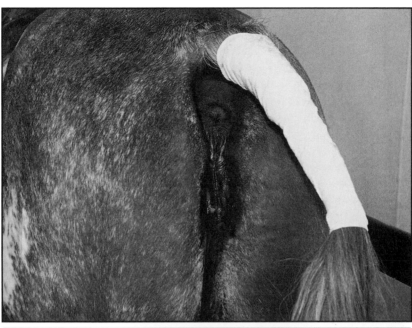

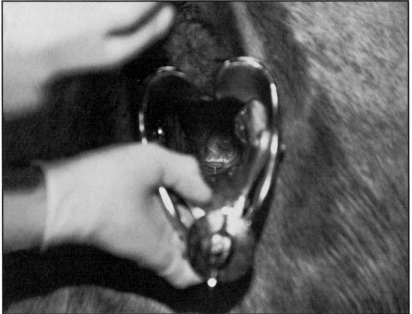

The mare's tail should be wrapped and the perineum scrubbed and rinsed prior to vaginal speculum examination. Note the relaxed cervix, characterisitc of estrus.

The mare's tail is wrapped and the vulva and perineum scrubbed and rinsed repeatedly using a nonirritating soap such as 2 percent Chlorhexidine solution. Surgical gloves are donned. A sterile vaginal speculum is inserted into the vestibule. The speculum is slowly advanced into the vagina. Temporary resistance occurs if there is an intact *vulvovaginal fold*. This is a flap of wrinkled vaginal mucosa that runs horizontally across the vaginal floor behind the opening of the urethra. The vulvovaginal fold prevents urine from running back into the vagina and is a major defense against bacterial contamination. It is weakened or lost in some mares with infertility.

In the maiden mare, a persistent hymen may be found at the junction between the vestibule and vaginal canal. A completely intact or imperforate hymen blocks the drainage of cervical secrections and will lead to uterine infection. Typically, the hymen is only partly intact and consists of one or two transverse bands across the vaginal entrance. These bands may interfere with breeding or cause heavy bleeding with the first mating. Your veterinarian can break these transverese bands by gentle finger traction. A thick or imperforate hymen will need to be surgically divided. These procedures should be done at least 2 to 3 weeks before breeding.

The color of the vaginal lining and appearance of the cervix correlate with the phase of the *estrous cycle* (discussed below). In *estrous*, the vaginal mucosa is a glistening pink to red color and the cervix is relaxed and short. In *diestrus*, the mucosa is gray to pale-white and dry. The cervix is firm, tightly constricted, high and located well off the vaginal floor.

A pool of white or gray watery fluid on the floor of the vagina or seen coming from the cervix indicates infection and the need to evaluate the uterus by culture, cytology and endometrial biopsy. This involoves passing instruments into the uterine cavity. When instruments are passed into the uterus there is a risk of introducing bacteria and causeing infection. Accordingly, procedulres that invade the uterus are not routinely done during a BSE, but may be recommended by your veterinarian if the mare is infertile and barren, is a repeat breeder (i.e., one who has failed to become pregnant on three consecutive heat cycles), has a history of abortion or early pregnancy loss, or has abnormal heat cycles. In addition, endometrial biopsy may be requested as part of a mare pre-purchase agreement. How these procedures are performed is explained in ENDOMETRITIS, chapter 8.

Yellow fluid with a urea-like odor pooling around the cervix is urine. It is indicative of *urovagina*, another significant cause of infertility.

Direct palpation of the vagina and cervix is performed after completing the vaginal speculum exam. The labia are spread and a gloved hand is inserted through the vestibule into the vagina. The degree of reisitance of the vulvovaginal sphincter is noted. The canal and vault are explored for adhesions, cysts and other abnormalities. The cervix is palpated for deformities associated with old injuries. In broodmares (but not maiden mares) the cervix is gently dilated and a finger inserted into the canal to explore for lacerations and adhesions.

At the conclusion of the vaginal examination, it is important to evactuate residual air from the uterus. This is accomplished by compressing the utuerus through the wall of the rectum.

THE REPRODUCTIVE CYCLE

NATURAL BREEDING SEASON

Mares are seasonal breeders. They exhibit sexual activity only during the breeding season, at which time they come into and out of heat at regular intervals.

The physiologic or natural breeding season should be distinguished from the man-made or operational breeding season discussed below.

The natural breeding season is determined by a number of factors, including latitude, length of daylight, temperature, nutrition, rainfall and climate. After a period of ovarian and reproductive inactivity called the winter anestrus, the natural reproductive season in the Northern Hemisphere begins in April and continues through October. In high northern latitudes it is shorter, while in deep southern latitudes it is longer and may not be accompanied by winter anestrus. The adaptive advantage of seasonal breeding is that foaling will always occur in spring, when weather and forage conditions are most favorable for raising a foal.

The length of daylight has a direct bearing on ovarian activity, with increasing hours of daylight associated with awakening of sexual activity. This effect is mediated through connections between the retina, optic nerves, pineal gland, hypothalamus, pituitary and ovaries. The retina is sensitive to the amount of light received during the day. This information is transmitted via the optic nerves to the pineal gland. The pineal gland secretes a hormone called melatonin. The amount secreted is dependent upon the length of day and night. Short days and long nights are associated with greater amounts of melatonin.

Melatonin suppresses the release of the gonadotropin-releasing hormone (GnRH), which is manufactured in the hypothalamus. The principal function of GnRH is to cause the pituitary gland to produce and release the *gonadotropins* that activate the reproductive cycle. Thus when days are short and nights are long, GnRH production is suppressed and gonadotropin output is negligible—effectively shutting down the ovaries. This explains why the mare remains in a state of sexual inactivity during the winter.

Although GnRH is barely detectable during deep winter anestrus, after 2 to 3 weeks of exposure to increasing periods of daylight, pulses of GnRH are released from the hypothalamus and levels begin to rise. As the photoperiod increases, the number of GnRH pulses also increases. The frequency of these pulses determines which hormones will be released from the pituitary gland.

Two to four pulses per day trigger the release of the follicle-stimulating hormone (FSH). FSH wakes up the ovaries and begins the process of follicle enlargement. This process occurs during the transition from winter to spring in late February and March.

When pulses occur every 2 hours, the pituitary releases luteinizing hormone (LH). LH causes a preovulatory follicle to mature and release an egg. This corresponds to the first ovulation of the natural breeding season.

Note that FSH and LH follow a distinct seasonal profile, with plasma LH rising more slowly than FSH in the spring and dropping more quickly than FSH in the fall. The discrepancies between the concentrations of FSH and LH in early spring and late autumn account for why some mares develop *anovulatory follicles* during spring and fall.

December 22, the winter solstice, is the day with the shortest period of daylight; 70 percent of mares are in deep anestrus at this time, and the number increases to about 85 percent by mid-January. In March, some of these mares begin to develop ovarian follicles. The first cycles might not be accompanied by ovulation; but by the middle of April, with the increasing number of daylight hours, more sun, warmer temperatures and green grass, the majority of mares experience the first ovulation of the year.

OPERATIONAL BREEDING SEASON

Thoroughbred racing associations and most breed registries have designated January 1 as the universal birth date. In effect, all foals born during the year automatically become yearlings on January 1 of the following year. (In the Southern Hemisphere, the universal birth date is August 1.) Thus the actual age of foals born during the year could vary by as many as 365 days.

The effect of the universal birth date is to make foals born early in the year more valuable than foals born later in the year, since early foals are more developed as two-year-olds and more likely to excel in competition.

A mare has a relatively long gestation (nearly one year). This means that if she is going to produce foals early in the year, breeding must begin during the winter months of subfertility. By manipulating the mare's natural breeding season using an artificial light program, it is possible to create a man-made or *operational* breeding season that begins on February 15 and ends on July 15. The operational breeding season thus overlaps the physiologic breeding season.

Artificial Light Program. By exposing a mare to an artificially lengthened photoperiod using natural light supplemented with artificial light, winter anestrus can be shortened and follicular activity advanced by about 2 months.

During the transition period between winter anestrus and the date of first ovulation, mares normally develop ovarian follicles that do not ovulate. These anovulatory follicles are difficult to distinguish from those destined to ovulate spontaneously. This complicates the question of when to breed. The use of artificial light advances only the date of follicle development; it does not advance the date of ovulation itself. Therefore breeding objectives frequently require the use of *human chorionic gonadotropin* (HCG) to induce ovulation in these transitional follicles (see INDUCING OVULATION in chapter 2, "Abnormal Estrous Cycles").

It takes at least 60 days to induce follicular activity with artificial light. If the breeding season is to start in early February, lighting should be initiated between November 15 and December 1. Experience indicates that the mare should be exposed to 16 hours of continuous light (natural plus artificial) per day until natural daylight approximates 16 hours (early June in the Northern Hemisphere). The schedule must be consistent. An irregular schedule will not advance the breeding season. The best way to guarantee a regular schedule is to equip the system with an automatic timer that comes on before dark and goes off after the mare has been exposed to the requisite number of hours of continuous light.

The entire 16 hours can be delivered starting with the first day of the program. As an alternative, light can be added at the rate of 30 minutes a week until 16 hours are achieved. The fixed 16-hour program is preferred since it has been studied extensively and appears to provide the most consistent results.

Studies show that 12 foot-candles of light in a 12×12–foot box stall is sufficient to extend the photoperiod. A 200-watt incandescent light bulb or a 400-watt fluorescent light bulb should provide at least 12 foot-candles of light, assuming that the source is located within the stall. For pens and paddocks containing incandescent, mercury vapor, sodium or quartz lights, light intensity should be measured by a lighting specialist to ensure adequate exposure in all areas of the enclosure.

THE ESTROUS CYCLE (HEAT CYCLE)

When a filly reaches puberty, she becomes sexually mature and begins to produce eggs. This generally occurs between 12 and 24 months of age. Mares younger than 2 years of age, however, are physically immature and ordinarily are not bred until they are at least 3 years old. Fillies over 10 months of age should be separated from stallions to prevent accidental pregnancy.

During the breeding season, the mare comes into and out of heat at regular intervals. One such interval is called an *estrous cycle*. It can be defined as the period from one ovulation to the next. The length of a normal estrous cycle is 21 to 23 days and consists of 14 days of diestrus, and 7 days of estrus including ovulation. The normal estrous cycle can vary by 1 or 2 days, especially at the beginning and end of the breeding season.

ESTRUS (THE FOLLICULAR PHASE)

Note that there may be confusion in spelling and terminology because the first part of the estrous cycle is called *estrus*.

Estrus is the period of sexual receptivity. It is relatively short. Early in the breeding season, it lasts 6 to 8 days and by mid-season decreases to about 4 days. A mare in estrus is actively interested in mating and displays receptive behavior in the presence of a stallion (see chapter 4, "Determining When to Breed"). Receptive behavior continues throughout estrus and ovulation. Ovulation occurs at the end of estrus, about 24 hours before the mare goes out of heat.

The hormonal effects that govern estrus and ovulation are complex. In brief, the pituitary gland releases follicle-stimulating hormone (FSH), which activates the ovaries and causes many follicles to grow and produce large amounts of

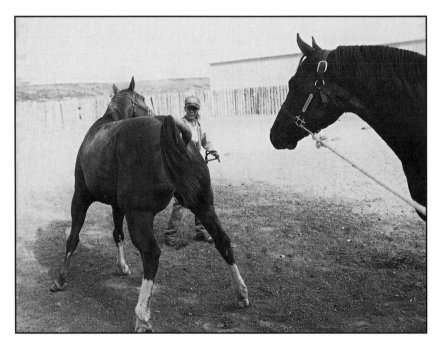

A receptive mare demonstrating the typical breeding stance of _estrus_.

estrogen. Estrogen prepares the reproductive tract for mating and fertilization. It is also responsible for the typical behavior of the mare in heat.

As follicles approach maturity, the pituitary gland releases a second hormone called luteinizing hormone. LH causes one or two follicles to become dominant. When a dominant follicle reaches about 45 mm, LH induces it to ovulate.

DIESTRUS (THE LUTEAL PHASE)

Diestrus is a period of sexual inactivity that begins 24 to 48 hours after ovulation and lasts for 14 days. It is marked by an abrupt change in behavior in which the mare refuses the stallion and exhibits her displeasure by laying back her ears, wheeling, squealing, kicking and occasionally biting and pawing.

Eight hours after ovulation, the ovulatory follicle fills with blood and serum. This produces a soft, mushy structure called the corpus hemorrhagicum (CH). During the next 2 to 3 days, the CH is replaced by a yellow mass within the ovary called the corpus luteum (CL). Five days after ovulation the CL is fully mature.

The most important function of the CL is to produce the hormone progesterone. Progesterone prepares the lining of the uterus to receive, support and maintain the embryo. If the CL fails in this task, the embryo will be lost. Progesterone is also responsible for the unfriendly behavior of the mare toward the stallion that occurs in diestrus.

The laid-back ears and unfriendly behavior of the diestrus mare.

If after ovulation the mare does not become pregnant, the CL remains active for only 12 to 14 days. It then undergoes rapid regression and ceases to function, this process is called *luteolysis*. As the CL disappears, the serum progesterone drops accordingly. This is followed in 1 to 2 days by a return to estrus.

Luteolysis is brought about by a hormone called prostaglandin PGF2α, which is produced by the mare's endometrium. On the 13th day of diestrus, pulses of PGF2α are released from the lining of the uterus into the bloodstream and are carried to the mare's ovary. PGF2α acts directly on the CL to initiate luteolysis.

If after breeding the mare does become pregnant, the uterus recognizes the embryo and inhibits the release of PGF2α. This ensures that the CL does not undergo luteolysis and that the pregnancy will continue.

The process by which the mare's uterus recognizes pregnancy is unusual. Newly fertilized embryos arrive in the uterus at about 6 days postovulation. Instead of implanting in the wall of the uterus, the embryos become nomadic, passing back and forth between the uterine horns and contacting the entire surface of the endometrium before finally implanting at the base of one of the horns at about 16 days' gestation. This wandering back and forth across the endometrium activates the pregnancy-recognition system and inhibits the uterus from releasing PGF2α.

At 70 to 90 days of gestation, the CL is no longer required. The placenta now assumes the role of producing the pregnancy hormone. At this time it would be possible to remove a mare's ovaries without causing her to abort.

CARE OF THE BROODMARE

Among mares bred annually in the United States, 40 to 50 percent do not produce live foals. The principal reasons for such low reproductive efficiency are poor breeding preparation, uterine infections and abnormal estrous cycles.

Good preparation involves maintaining an excellent state of health and nutrition through vaccinations, a parasite-control program, preventive dentistry and good foot care.

FEEDING AND NUTRITION

The minimum daily requirements for the broodmare are the same as those for the mature horse at maintenance, shown in Table I, appendix. Only the best-quality roughage and horse feeds should be used.

Pregnancy conception rates are influenced by body condition. Mares who are excessively fat or thin do not conceive as readily. These mares should be placed on an appropriate diet to achieve better body condition. If a mare is just moderately overweight, however, she should not be placed on a weight-losing diet, since this may compromise her nutrition. Furthermore, there is no evidence that a moderate degree of obesity interferes with fertility. The extremely thin mare is often suffering from a heavy burden of intestinal parasites and should be treated.

Ideally, a mare should be slightly lean coming out of winter anestrus and just starting to gain weight. In fact, if the mare is actually lean, you can increase the calories in her daily ration by adding 2 to 3 pounds of grain per day. Begin 4 to 6 weeks before breeding. This process is called *flushing*. Studies suggest that flushing improves a mare's reproductive performance. However, if the mare is somewhat fat (as is all too common), do not compound the problem by adding grain to her diet.

A mare coming directly from the racetrack or a strenuous training program may need a let-down period of up to 60 days to establish regular estrous cycles. Mares given anabolic steroids generally do not cycle for at least 6 months.

Vitamin and mineral supplements are not required unless the broodmare has been consuming a substandard ration or is severely run-down from a recent pregnancy. Commercial horse feeds and most hays and grasses contain vitamin and mineral concentrations that meet or exceed the horse's daily needs. Green forages (pasture or hay) are good sources of vitamins A and E. These are the only vitamins not synthesized by the horse and therefore the only vitamins that must be provided by the diet. Vitamin E deficiency (with the possible exception of nutritional myopathy) has not been a problem in horses.

Vitamin A deficiency could possibly occur if a mare were unable to consume green forage or another dietary source of vitamin A for several months.

A broodmare does not have increased needs for calcium and phosphorus. Nor will micromineral deficiencies occur if the mare is allowed unrestricted access to trace-mineralized salt. In selenium-deficient areas, use trace-mineralized salt containing higher levels of selenium.

VACCINATIONS

A recommended immunization schedule is shown in Table II, appendix. All vaccinations should be current to protect the health of the mare and her future foal. Foals are at increased risk for certain infections that develop shortly after birth. Protection is afforded by high levels of maternal antibodies in the colostrum. High levels are maintained by a series of basic vaccinations followed by regular boosters at specified intervals.

Table III in the appendix shows additional vaccinations recommended for horses in endemic and high-risk areas. If you are shipping your mare some distance for breeding, additional vaccinations may be indicated. To learn if a disease is endemic, consult a veterinarian or state extension agent in that area.

Regular deworming prevents infestations that can impair fertility.

DEWORMING

All horses are infested with intestinal parasites. A heavy worm infestation, particularly with large and small *strongyles*, can severely impair a mare's health and reproductive efficiency. Since it is impossible to prevent worms, the goal of regular deworming is to reduce the burden of parasites and maintain low levels of infestation.

A number of deworming agents are available (see INTESTINAL PARASITES in chapter 7, "Pregnancy"). Each has certain advantages when applied to variables such as climate, geographic location, number and concentration of horses, convenience, expense and history of problems in the past. It is important to include your veterinarian in the development of a program that best meets your specific needs.

If your mare has been on a regular deworming program, no additional deworming is required in preparation for breeding.

PREVENTIVE DENTISTRY

Hay and grain that is not broken down by the grinding action of healthy teeth is incompletely digested. Loss of nutritional value in feed is particularly detrimental in the broodmare, because growth and development of the foal is highly dependent on the nutritional condition of its mother.

Young to middle-age horses develop extremely sharp edges and points on their molars, while older horses may develop loose teeth and infected sockets. These dental problems cause mouth pain, interfere with chewing and lead to abnormal patterns of mastication.

Veterinary examination and floating (filing) the teeth is recommended on a yearly basis for all horses over 2 years of age. The purpose of floating is to file down the sharp edges and points on the upper and lower premolars and molars. Filing is done with a long-handled rasp. The procedure is not painful and can be done with minimal restraint.

FOOT CARE

A program of daily inspection and foot cleaning, regular exercise and the application of horseshoes for those horses who require them will prevent many foot problems.

The hooves of the shod mare should be trimmed every 5 to 8 weeks. If the mare is unshod, trimming may be required less often.

Before bringing the mare to the breeding shed, be sure to remove her back shoes to reduce the likelihood of serious injury if she kicks the stallion.

ABNORMAL ESTROUS CYCLES

Mares with abnormal estrous cycles either: (a) do not exhibit heat, (b) exhibit persistent heat, or (c) come into and out of heat at irregular intervals (the heat periods being shorter or longer than the average 21 to 23 days).

Anestrus can be defined as absence of heat resulting from inactive ovaries. Anestrus mares do not cycle. This is the normal condition during winter.

Absence of heat may also occur because of prolonged diestrus. Pregnancy, a persistent corpus luteum, pseudopregnancy and ovulation in diestrus are all causes of prolonged diestrus.

Persistent estrus is caused by *anovulatory follicles* that produce excess estrogen. Rarely, persistent heat is caused by an estrogen-producing granulosa-thecal cell tumor of the ovary.

Irregular estrous cycles are seen with split estrus, short-cycling, ovulation in diestrus and transitional heat periods.

Diagnostic studies that help to determine whether a mare has a normal or abnormal estrous cycle include rectal palpation, ultrasound of the ovaries and measurement of serum progesterone. Serum gonadotropin measurements and genetic testing may be indicated if further investigation is necessary.

ANESTRUS

Anestrus is lack of estrus or failure to cycle. It should be distinguished from diestrus, the phase of the heat cycle in which the mare is not receptive to the stallion. Diestrus lasts 14 days and is followed by a return to heat. When the length of diestrus exceeds 14 days, the mare is said to be in a state of prolonged diestrus, also referred to as a prolonged luteal phase, discussed below.

Anestrus and prolonged diestrus are different conditions but may appear similar because both are characterized by a lengthy period in which the mare does not show signs of heat. These conditions can be distinguished by rectal or ultrasound examinations, as well as serum progesterone levels drawn at 10-day intervals. Ovaries that are smooth and small are typical of anestrus. Ovaries that contain follicular cysts and/or a corpus luteum are hormonally active and typical of diestrus. Low progesterone levels occur during anestrus and estrus; high levels occur during diestrus.

Seasonal anestrus is the most common cause of ovarian inactivity during the winter months.

Ovarian tumors and sex chromosome abnormalities are unusual causes of anestrus (see chapter 8, "Mare Infertility"). Tumors of the hypothalamus and pituitary gland can cause anestrus by interfering with the production of FSH. These tumors are rare.

The following are the common causes of anestrus.

A severely malnourished mare with debilitation anestrus.

DEBILITATION ANESTRUS

Malnourished, overworked and stressed mares frequently do not come into heat. Mares are often stressed by racing and hard training, as well as by the transport required to compete in the various events. A debilitated state accompanied by weight loss is usually caused by chronic disease; a heavy burden of intestinal parasites; or a diet deficient in energy, protein and essential nutrients.

Rectal palpation and ultrasound reveal small, hard, inactive ovaries. The uterus is flaccid. The mare shows no estrus signs when teased by the stallion. Correcting the cause of the debilitation may correct the hormone problem and allow the mare to resume ovarian activity.

POSTPARTUM (LACTATION) ANESTRUS

The majority of postpartum mares come into heat 4 to 14 days after foaling (the average being 7 to 10 days). However, 5 percent of postpartum mares fail to reestablish normal estrous cycles for up to 3 months. These mares remain barren at a time when many farm managers would like to see their mares back in foal (see BREEDING ON THE FOAL HEAT in chapter 4).

A frequently used but inaccurate term for postpartum anestrus is lactation anestrus. It is now known that absence of heat is caused by either ovarian inactivity or a prolonged luteal phase after the foal heat ovulation, but not by the presence of lactation.

Ovarian inactivity tends to affect mares during the months of seasonal transition: January, February and March. Mares who foal during these months are more likely to exhibit postpartum anestrus because of the occurrence of anovulatory follicles. Old-age mares and thin mares with a poor body-condition score (see appendix) are more likely to suffer from postpartum anestrus regardless of season.

Treatment: Postpartum anestrus caused by a prolonged luteal phase responds to prostaglandin therapy as described for Persistent Corpus Luteum. The mare should return to estrus about 4 days after receiving an injection.

Postpartum anestrus caused by ovarian inactivity may persist for 1 to 3 months. It usually disappears in late spring. To minimize the effects of postpartum anestrus, maintain optimal body condition during pregnancy and expose the mare to 15 hours of continuous light beginning 2 months before foaling (see chapter 1).

STEROID-INDUCED ANESTRUS

Performance mares given anabolic steroids containing derivatives of testosterone often have small, inactive ovaries that fail to develop follicles. In low doses, anabolic steroids suppress heat behavior but not the estrous cycle. In higher doses, they shut down the estrous cycle entirely. Negative effects can persist for up to 6 months after the steroids are stopped and may even be permanent. Accordingly, performance-enhancing drugs should not be used in mares for whom breeding is planned.

PROLONGED DIESTRUS

Diestrus is the period of the estrous cycle associated with the presence of the corpus luteum. The corpus luteum (CL) manufactures progesterone. Progesterone levels are high during diestrus and fall rapidly at the end as the CL regresses and disappears. Diestrus is also referred to as the luteal phase of the estrous cycle, or the period of luteal activity.

All mares, especially those who are barren, maiden or lactating, can fail to exhibit regular heat cycles and remain in diestrus for prolonged periods as determined by an elevated serum progesterone.

Note that diestrus begins immediately after ovulation and lasts 14 days. The high progesterone level prevents the mare from returning to heat. As long as the CL continues to produce progesterone, the mare remains in diestrus. Accordingly, any condition that prolongs the life of the CL will stop the mare from cycling.

After a mare has been in diestrus for 13 days, pulses of prostaglandin PGF2α are released from the lining of her uterus and carried to the ovary. The PGF2α acts directly on the CL to cause it to regress in size and disappear. This process is called luteolysis. With luteolysis, the progesterone level drops rapidly. The mare returns to heat in 1 to 2 days.

Any disorder of the luteolytic mechanism that prolongs the diestrus phase of the estrous cycle is a serious problem to a breeding program because it prevents breeding. The common causes of prolonged luteal activity are a persistent corpus luteum (including pregnancy), pseudopregnancy, ovulation during diestrus, and postpartum (lactation) anestrus (discussed above).

PERSISTENT CORPUS LUTEUM

Pregnancy is the most common cause of a persistent corpus luteum. As described in DIESTRUS (THE LUTEAL PHASE) in chapter 1, the pregnant uterus prevents the release of the luteolytic hormone PGF2α and this ensures that the pregnancy survives.

The endometrium also may not manufacture and release PGF2α for reasons other than pregnancy. A uterine infection, for example, can destroy the lining of the uterus to such an extent that the residual endometrial glands are too few in number to produce sufficient amounts of prostaglandin to affect the CL. Thus chronic endometritus is a cause of a persistent CL.

A normal endometrium can fail to release PGF2α when the uterus is disturbed by examinations or procedures that allow air and contaminants to enter and irritate the lining.

Spontaneous persistence of a CL without an apparent cause has been described but is believed to be rare.

The diagnosis of a persistent CL is based on the history of a mare who does not go into estrus during the breeding season, and the findings on rectal palpation of a normal-sized uterus with firm tone and a long-closed cervix. The CL is buried within the ovary and usually cannot be palpated. *Ultrasonography* visualizes the CL in the majority of cases. An elevated serum progesterone after the 14th day of diestrus is also diagnostic of a persistent CL.

Treatment: A single injection of PGF2α will cause *luteolysis* in more than 90 percent of cases. The mare should return to estrus within 4 days and have a normal heat cycle. The length of estrus and the exact day of ovulation will depend upon how large the preovulatory follicles were at the time of the injection.

OVULATION IN DIESTRUS

Ovulation in diestrus is believed to account for most cases of prolonged diestrus.

The mare has a unique ability among domestic animals to ovulate more than one follicle per cycle. Double ovulation is noted in about 10 percent of cycles. The first ovulation occurs at the normal time: 24 hours before the end of estrus. The second ovulation occurs 2 to 15 days after the first ovulation. This means that the second ovulation takes place during diestrus. Although the second ovulation is fertile, it is not accompanied by signs of heat. Twin pregnancy may result from a double ovulation. If the two ovulations are spaced close together, both eggs may be fertilized from a single mating. When the ovulations are spaced several days apart, it usually takes a second breeding to fertilize the second egg. Twin pregnancy is discussed in chapter 9, "Abortion."

A diestrus ovulation early in diestrus will not affect the length of the estrous cycle. Both CLs will mature and respond to *luteolysis* at approximately the same time. As a consequence, both will undergo luteolysis at the same time. Diestrus lasts the standard 14 days, and the mare returns to heat.

However, a second ovulation *late* in diestrus (more than 10 days after the first ovulation) will prolong luteal activity and cause the mare to stop cycling. This is because the second CL is less than 5 days old at day 13. It is still too young and immature to respond to the effects of PGF2α released after the first ovulation. Thus the second CL, having escaped luteolysis, will persist for a variable and unknown length of time.

If a mare has one diestrus cycle, she is apt to have several more, resulting in long periods of absent heat behavior.

Treatment: The veterinary diagnosis of diestrus ovulation in the nonpregnant mare is based on ultrasound findings showing a second ovarian follicle and/or the presence of more than one CL. It is important to determine that the mare is not pregnant, since prostaglandin treatment could result in abortion. Plasma progesterone concentrations typically show high levels (greater than 2 ng/mL) that persist beyond 14 days.

A single injection of PGF2α will cause luteolysis of the second CL, provided that it is at least 5 days old. If the CL is immature, it will not respond to luteolysis and the mare will not return to estrus. If the mare does not return to heat, a second injection is administered 7 to 10 days later.

The mare who repeatedly ovulates diestrus follicles in consecutive estrous cycles and fails to show signs of heat may need to have her follicular activity temporarily suppressed in order to restart an orderly cycle. Your veterinarian can accomplish this by administering 100 to 200 mg of oral progesterone daily for 10 to 14 days, or by giving a combination of 150 mg of progesterone and 10 mg of estrogen daily for 10 days.

PSEUDOPREGNANCY

Pseudopregnancy in the mare is not well understood. Most cases occur in mares who have been unsuccessfully bred; however, pseudopregnancy has also been described in non-bred mares as well. The reasons for this are unknown.

The pseudopregnant mare exhibits all the signs of pregnancy including lack of estrus; tense uterine tone; a closed, firm cervix; and a pale, dry vagina. However, an ultrasound exam finds that the mare is not pregnant.

Most cases of pseudopregnancy appear to be the result of an actual pregnancy that ended with death and loss of the fetus at an early stage of *gestation*. The CL associated with the pregnancy does not regress as one might expect with the termination of the pregnancy. A high plasma progesterone concentration is maintained for several weeks.

How long the CL will persist and prevent the mare from cycling depends on when the conceptus was lost. When embryonic death occurs after the 11th day but before the 36th day of pregnancy, remnants of fetal tissue may remain in contact with the endometrium for a variable period. The corpus luteum persists until all fetal tissue is expelled or reabsorbed. In the majority of cases, this takes about 1 to 3 weeks, after which the mare returns to estrus and cycles regularly.

However, when abortion occurs after the 37th day, luteal activity persists for a much longer time. At this exact time, specialized fetal glands develop from the placenta and interlock with the endometrium. These glands, called *endometrial cups*, are independent of the presence of the fetus. They persist in the wall of the uterus until the 110th to 120th day of gestation, when they cease to function. The endometrial cups produce equine chorionic gonadotropin (eCG). The function of eCG is to ensure that the CL remains active and continues to produce the progesterone necessary to support the pregnancy. Accordingly, the mare who aborts on day 60 of gestation, for example, will not return to estrus and ovulate for 50 to 60 days.

Treatment: A pseudopregnancy can be terminated with one or more injections of PGF2α, depending on the duration of the pseudopregnancy.

PERSISTENT ESTRUS

ANOVULATORY FOLLICLES ASSOCIATED WITH TRANSITIONAL HEAT CYCLES

The production of egg follicles that do not ovulate occurs commonly at the beginning and end of the natural breeding season.

Ultrasound of the ovaries during the spring transition reveals many immature follicles 10 mm to 20 mm in size, and occasionally some follicles that are larger. Because of the low concentrations of luteinizing hormone (LH) early in the breeding season, these immature follicles are not capable of ovulating. Eventually they collapse and are replaced by new follicles. Several such crops may occur before the mare finally produces a dominant follicle (35 to 40 mm) at a time that coincides with an LH surge. This timing results in the first ovulation of the breeding season and is followed by regular estrous cycles.

Estrogen produced by the anovulatory follicles is responsible for the prolonged and occasionally inconsistent heat behavior of the mare. The mare may come into and go out of heat many times in a short period, or remain in heat for weeks on end. While in heat, the mare is receptive to the stallion and will usually stand for breeding. In the spring, this behavior may last for more than 60 days.

Treatment: Spring transitional estrus is a physiologic event that disappears in May or June with exposure to longer periods of daylight.

Treatment can be employed to expedite early breeding. If the mare has a transitional follicle of at least 35 mm, ovulation can be induced with human chorionic gonadotropin (HCG) (see INDUCING OVULATION later in this chapter). This is successful in about 70 percent of cases. If successful, ovulation occurs within 48 hours. The mare should be bred immediately in anticipation of ovulation.

When multiple small anovulatory follicles are present, these follicles should be suppressed to allow the next generation of follicles to develop. This best way to accomplish this is by using a drug protocol that combines 150 mg of progesterone with 10 mg of estradiol. These two drugs are given by intramuscular injection daily for 10 days. In some protocols, a single injection of the prostaglandin Lutalyse (1 mg/100 lb. weight of horse) is added on the 10th day. As soon as a preovulatory follicle reaches 35 mm, ovulation is induced with HCG.

Anovulatory follicles in the months of March and April can be avoided by advancing the breeding season using an artificial light program (see chapter 1). Frequent teasing by the stallion is also helpful in establishing regular estrous cycles.

ANOVULATORY HEMORRHAGIC FOLLICLE (AUTUMN FOLLICLE)

On rare occasions (most often in autumn) a follicle fails to rupture and fills with blood. Over a period of days, the follicle can become quite large, often 60 to 90 mm in diameter. The follicle on ultrasound is seen to be an encapsulated mass on the ovary containing a gelatinous fluid. A rim of corpus luteum inside the capsule may be present. Unlike the waves of follicles that occur in spring, hemorrhagic follicles do not produce much estrogen.

Anovulatory hemorrhagic follicles are a reflection of the hormonal patterns present during seasonal transition, when there is inadequate output of pituitary gonadotropins. These follicles generally regress and disappear in about one month.

IRREGULAR ESTROUS CYCLES

An irregular estrous cycle is one in which heat occurs more frequently, or less frequently, than every 21 to 23 days. Such cycles are common during the spring transition and disappear after the first ovulation of the breeding season. Cycles longer than 23 days are characteristic of ovulation in diestrus (discussed above).

Two conditions associated with short heat intervals are short-cycling and split estrus.

SHORT-CYCLING

When a mare comes back into heat in less than 21 days, she is said to be short-cycling. Short-cycling is caused by early release of prostaglandin PGF2α before the 14th day of diestrus. This results in premature regression of the CL. Thus the short cycle is due to shortening of diestrus.

The CL cannot undergo luteolysis until the 5th day after ovulation. With the shortest possible cycle (5 days of estrus plus 5 days of diestrus), the mare could come into heat as often as every 10 days. In practice, short-cycling mares usually come into heat every 14 to 16 days.

Uterine infection (endometritis) is the principal cause of early release of PGF2α. Recall that endometritis is also a cause of delayed or absent release of PGF2α (see PERSISTENT CORPUS LUTEUM earlier in this chapter). Accordingly, when the uterus becomes infected, the mare is likely to have an irregular estrous cycle—but whether it will be shorter or longer than normal is difficult to predict.

PGF2α can also be released when the uterus is palpated, when the cervix is dilated, when the uterus is infused with antibiotics or when an endometrial biopsy is taken. Here again, the effects on the prostaglandin release mechanism are variable. There may be no effect, resulting in a normal cycle; prostaglandin release may be immediate, resulting in a short cycle; or release may be inhibited, resulting in a persistent corpus luteum and a long cycle.

Treatment: The short-cycling mare should be investigated and appropriately treated as described in chapter 8, "Mare Infertility."

SPLIT ESTRUS

In this condition, the mare comes into heat for several days, then passes into diestrus for 1 to 2 days, and finally returns to heat for several more days. Ovulation generally occurs during the second heat period. The hormonal basis for split heat is unknown. The condition appears to be more common early in the breeding season during the spring transition.

Split estrus presents problems because the mare is likely to be bred on the first heat period. This is followed by a period of diestrus, in which the mare rejects the stallion. It is assumed that heat is over. Breeding is stopped and the mare does not become pregnant.

Treatment: Split heat can be suspected when a mare fails to become pregnant after breeding on one or more cycles. Teasing the mare for several days after she goes out of heat should identify the second heat period. As an alternative, rectal palpation and/or ultrasonography will show that the mare did not ovulate. If breeding is continued through the second heat period, conception is likely.

HORMONAL MANIPULATION OF THE ESTROUS CYCLE

The estrous cycle can be hormonally altered to bring a mare into or out of heat, and to cause her to ovulate at a specified time. These manipulations can increase breeding efficiency and are often necessary to make early season breeding possible and to time ovulation with artificial insemination.

The four hormones commonly used to manipulate the estrous cycle are prostaglandin, progesterone, HCG and estrogen.

PROSTAGLANDIN

Prostaglandin F2α, known as *PGF2α* or *Prostin*, is widely used in mares to shorten or terminate a state of luteal activity in order to induce labor or abortion. Prostaglandin should not be used in pregnant mares except for the intended purposes of inducing labor or bringing about abortion.

The two commonly used prostaglandin products are Equimate™ (fluprostenol sodium) and Lutalyse™ (dinoprost tromethamine). Lutalyse is given as a single injection at an average dose of 1 mg/100 lb. horse weight (1 to 2 mL of the dinoprost tromethamine solution containing 5 mg/mL).

The average dose of Equimate is 250 micrograms (5 mL). For synchronizing ovulation, the two-injection method is recommended (see below). Note that prostaglandin dosages vary with the products and purposes for which they are used.

Prostaglandin preparations can be given by the intravenous, intramuscular or intrauterine route. All are equally effective. The IM route, which is used most often, has a lower incidence of side effects (10 percent) than does the IV route. Side effects, which may last for 45 minutes, include profuse sweating, defecation and transient abdominal discomfort. When the drug is given intravenously, side effects include incoordination, swaying, staggering and lying down. Serious drug reactions are rare.

All PGF2α preparations are easily absorbed through unprotected skin and can produce toxic reactions. Pregnant women, women of child-bearing age who may be unaware of a pregnancy, asthmatics, and individuals with respiratory diseases should handle these products with extreme caution to avoid spillage and contact with the skin. If accidental contact occurs, the skin should be washed immediately with soap and water.

PROGESTERONE

Progesterone is available in oral and injectable form. Injectable progesterones in water are administered at a dose of 150 to 300 mg. The injections must be administered daily. Long-acting depo-progesterone *in oil* is not effective in suppressing estrus or maintaining pregnancy in the mare.

The only orally active synthetic progesterone effective in horses is the liquid preparation altrenogest (Regu-Mate™). The average daily dose is 0.044 mg/kg (22 mg, or 10 mL of stock solution for a 500 kg mare). Regu-Mate can be given in the back of the mouth with a dose syringe or added to the daily grain ration.

An important therapeutic effect of progesterone is to prevent and control the occurrence of heat. Estrus will not occur while the drug is being given. After the drug is withdrawn, estrus will occur in 2 to 5 days.

Progesterone is effective in suppressing heat in show and racing mares (see Prevention of Estrus, chapter 12), in advancing the first ovulation of the breeding season, in synchronizing ovulation for breeding, and in delaying ovulation on the first postpartum heat. It is also used to maintain pregnancy in mares who repeatedly abort.

Progesterone should not be given to a mare with a history of uterine infection because of the risk of exacerbating the infection and causing a relapse.

HUMAN CHORIONIC GONADOTROPIN

This hormone is manufactured by the placenta and excreted in the urine of pregnant women. HCG closely mimics the effects of equine luteinizing hormone, causing the mature follicle to rupture and release an egg. It is the agent of choice for inducing ovulation in estrus mares.

HCG is available in a generic label and also under the trade names A.P.L.™ and Follutein™. It can be given by intravenous, intramuscular or subcutaneous injection. The usual dose is 2000 to 3000 International Units (IU) given as a single injection.

HCG stimulates the production of antibodies that diminish its effectiveness. To maintain efficacy, it is recommended that no more than two injections be given during one year. When given during pregnancy before the 36th day, HCG has been known to cause early fetal loss.

ESTROGEN

The most potent naturally occurring estrogen, and the one used most often, is estradiol cyprionate (ECP). Note, however, that estrogen currently is not approved by the FDA for use in mares.

Natural estrogen plays an important role in protecting the mare's reproductive system from acquired infections. It is also important in the maintenance of pregnancy and in the induction of labor. Established therapeutic uses are limited to a relatively few conditions. They include synchronizing ovulation, treating anovulatory follicles associated with the transition period, and delaying ovulation on the first postpartum heat. Estrogen in these situations is usually given as a single injection (10 mg) in combination with progesterone.

Estrogen is known to cause birth malformations of the urinary system and to depress the fetal bone marrow. It should not be used in pregnancy.

SHORTENING THE ESTROUS CYCLE

As an aid in scheduling insemination or breeding, it is often beneficial to shorten a mare's estrous cycle and advance her date of breeding. Shortening the estrous cycle increases the frequency of estrus periods, which is beneficial when treating endometritis.

A prostaglandin injection given 5 or more days after ovulation shortens the diestrus phase of the estrous cycle and brings the mare into heat within 2 to 4 days. Most mares will ovulate 8 to 12 days after the injection. However, if the mare has not been out of heat for at least 5 days, the CL will be too immature to respond to the prostaglandin, and the injection will need to be repeated (see PERSISTENT CORPUS LUTEUM earlier in this chapter).

Note that only the diestrus interval is influenced by prostaglandin. The estrus interval (the first day of estrus to the day of ovulation) is not affected. Thus PGF2α will shorten the normal estrous cycle by no more than 5 to 9 days.

When a mare fails to become pregnant after having been bred, shortening diestrus allows a quick return to heat. Using this strategy, all bred mares are examined by ultrasound 15 to 18 days after ovulation. Mares who are not pregnant are given an injection and brought back to the stallion for rebreeding in a shortened period of time.

SYNCHRONIZING OVULATION

Synchronizing ovulation is the process by which the cycles of two or more mares are hormonally regulated for the purpose of causing all of the mares to come into estrus and ovulate close to the same time.

Synchronizing cycles makes it possible to schedule a number of mares for insemination with fresh or cooled transported semen using a single ejaculate. It also permits a stallion with a busy event schedule to breed several mares during a relatively short absence from competition. Synchronizing the ovulations of donor and recipient mares is a necessity for successful embryo transfer.

Because of the long and variable follicular phase of the estrous cycle, there are limitations on the accuracy with which ovulation can be synchronized. Using present techniques, ovulation can be synchronized consistently to within a range of about 4 days.

Synchronizing ovulation is a complex undertaking that requires veterinary expertise. Note that all mares in a given program are treated with the same drugs at the same times, regardless of the phase of their estrous cycles. The choice of drugs is influenced by a number of variables, including the time of year, state of ovarian activity, and postpartum or anestrus state of the mares.

In the most common protocol, mares are given two injections of prostaglandin 15 days apart, followed by an HCG injection 6 days after the second prostaglandin injection. Within 3 days after the administration of HCG, 75 percent of mares will ovulate.

Another protocol uses the administration of oral progesterone (Regu-Mate) daily for 15 consecutive days. Estrus will ensue within 2 to 5 days, and most mares will ovulate 8 to 15 days after the last day of treatment.

Also used is a combination of intramuscular progesterone (150 mg) and estradiol (10 mg), given for 10 consecutive days, followed by a prostaglandin injection on the 10th day. Ovulation occurs 10 to 12 days after the prostaglandin injection in about 80 percent of mares. The optional administration of HCG (2500 IU) when a 35 mm follicle is detected may further improve the synchronization results.

INDUCING OVULATION

The purpose of inducing ovulation is to cause a follicle to rupture and release an egg at a convenient time for breeding—either by live cover or artificial insemination. The ability to precisely time semen deposit with ovulation has the potential to enable conception with a single breeding or insemination.

Ovulation can be induced by an injection of HCG, provided that one dominant follicle is in the 40 mm range as determined by rectal palpation or ultrasound of the ovaries. Following intramuscular or subcutaneous injection, ovulation will occur in 24 to 48 hours (with an average of 36 hours).

When inducing ovulation for breeding with cooled transported semen, two options are available for coordinating the HCG injection with the availability of the semen. The first is to administer HCG on the expected day of insemination. However, if the shipment is delayed 1 or 2 days, ovulation could occur before the semen arrives. The second option is to wait for the semen to arrive, inseminate the mare, and then give the HCG. To ensure that semen will be present in the mare's oviducts during the critical period of fertility, a second insemination is done 12 hours after the HCG injection. For more information, see ORDERING THE SEMEN in chapter 6, "Artificial Insemination."

THE STALLION

SELECTING THE STALLION

The stallion contributes 50 percent to the hereditary makeup of the foal. Therefore it follows that mares should be bred only to a stallions with outstanding characteristics.

The ideal stallion would have, among other attributes, superior type and conformation, a pedigree with proven bloodlines, a successful performance record, a good disposition and a history of producing outstanding foals. In practice, few stallions possess all such outstanding attributes. The trick is to evaluate each stallion in terms of his relative strengths and weaknesses and then pick the stallion you feel is best for your breeding program.

A breeding stallion must possess good body conformation. It is simply not worthwhile breeding to an animal who is structurally unsound and has other serious weaknesses that may be hereditary. Furthermore when a horse is anatomically correct, he usually performs better in athletic competition. In addition, a well-conformed stallion has fewer faults and is less likely to sire foals with faults.

Type is a set of distinguishing features that gives a horse the look of the purebred. A stallion of superior type possesses an appearance close to the ideal, as described in the standard of his breed. Type is synonymous with beauty and eye appeal. It is a valuable quality when the purpose of breeding is to produce horses for show and pleasure.

The stallion's pedigree is a record of his lineage for several generations. A distinguished pedigree contains the names of one or more famous ancestors who have had a major import. The value of a pedigree is greatest when the outstanding ancestor is only one or two generations removed. A familiarity with bloodlines is important, because some bloodlines consistently produce outstanding individuals.

A performance record that includes a number of show wins increases a stallion's desirability. Many traits that enable a horse to do well in competition are heritable. Speed, for example, is believed to be about 50 percent heritable.

The stallion contributes 50 percent to the hereditary makeup of the foal. (Pictured is Zans Parity, a champion Quarter Horse owned by Walsh Quarter Horses, Montrose, Colorado. Photo: J Bar D Studios, Grand Junction, Colorado.)

When a stallion has been used at stud for several seasons, his production record will be a matter of great significance. If he has sired outstanding foals out of a number of different mares, you have a good indication of his merit as a producer. Such stallions are eagerly sought after and usually command high stud fees.

The stallion's temperament and disposition are important but usually do not carry as much weight as his other attributes. Many people feel that the mare plays the more important role in determining a foal's temperament. Moreover, temperament and disposition depend as much on socialization, training and prior handling as they do on heredity.

The stud fee, contract policies and costs of travel and maintenance are important financial considerations that can influence choice of stallion. The stallion owner normally charges a booking fee (which may be deducted from the stud fee) as well as the stud fee itself. In the event that the mare does not become pregnant, most stallion owners will not refund the stud fee but instead guarantee a return service at no charge. Guarantees ordinarily do not cover the cost of boarding and other expenses. Return-service guarantees usually last only through the next breeding season. All arrangements, including the signing of the stud contract, should be completed prior to the breeding season. If you are not comfortable with the assurances and guarantees of the stallion owner, consider using another stallion.

Finally, if you are planning to breed by artificial insemination using cooled transported semen, it is important to know how well the stallion's semen survives the cooling and thawing process. Many potent stallions produce semen that does not ship well. Even if the stallion is ideal in all other respects, if his semen does not ship well, he will not be useful for your purposes. For more information on what to look for, see chapter 6, "Artificial Insemination."

STALLION ANATOMY

The stallion has two testicles enclosed by the scrotum and located in the pubic region. The testicles produce sperm and also the male hormone testosterone, which is responsible for the secondary sex characteristics of the stallion. The normal adult testicle is about 3 to $5^1/_2$ inches in length and 2 to $3^1/_2$ inches in width. Sperm are produced throughout the year, but maximum testicular size and activity occur during the natural breeding season in May, June and July. Stallions with large testicles produce more sperm than stallions with small testicles.

The temperature within the scrotum is 2 to 3 degrees below body temperature. This lower temperature is necessary for sperm production.

The epididymis is a coiled tube resting on top of the testicle and connecting the testicle with the spermatic duct. The epididymis serves as a reservoir for sperm during their final days of maturation. Maturation takes 21 days. The sperm will die if they are not ejaculated shortly after they mature.

Sperm are transported via the spermatic duct up into the ampulla and thus to the urethra. The urethra is a long tube beginning at the bladder and ending at the tip of the penis. It serves as the passageway for both urine and semen. Also entering the urethra and mixing with the sperm are the secretions of the accessory sex glands, which are the vesicular glands, prostate and bulbourethral glands. These glands produce seminal fluid, which provides energy and protective buffers for the sperm. The combination of sperm and seminal fluid is called semen.

The penis is a cylindrical structure about 20 inches long when relaxed. During erection, spongy tissue fills with blood, increasing the length and diameter of the penis by about 50 percent. At the end of the penis is the bell-shaped glans. In the horse, the tip of the urethra projects slightly beyond the end of the glans. This extension is called the urethral process.

The glans of the penis contains additional erectile tissue, which engorges after the penis enters the vagina. This marked enlargement of the glans is called belling or flowering. When belled, the penis seats itself against the cervix so that the urethral process is coupled with the opening of the cervical canal. Semen is ejaculated directly into the uterus. If the penis does not bell, the stallion cannot ejaculate.

A discrepancy may exist between the length of the penis and depth of the mare's vagina. A stallion with an exceptionally short penis, for example, may be unable to seat the glans during belling. At the opposite extreme, a stallion with an exceptionally long penis may bruise or tear the mare's cervix or vaginal wall. Such injuries can be prevented by using a breeding roll, as described in COVERING THE MARE in chapter 5.

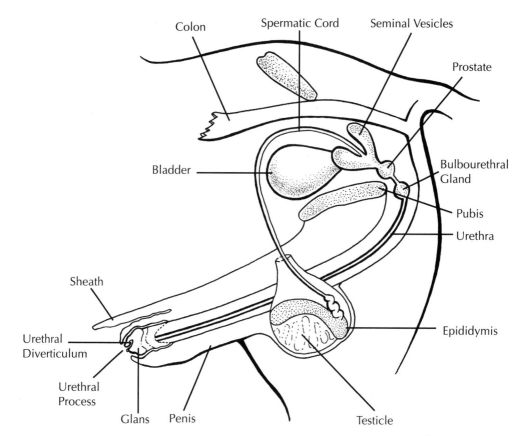

Stallion reproductive system.

Ejaculation is controlled by nerve pathways that begin in the brain and ultimately connect with the penis. During ejaculation, muscles surrounding the urethra and base of the penis contract six to eight times to expel the semen. Most of the sperm are present in the first four pulsations.

The sheath surrounding the penis is called the prepuce. The prepuce is a double layer of sliding skin. The internal layer contains sebaceous glands; its secretions, together with flaking skin cells, form a thick, waxy material called smegma. Smegma tends to collect in the folds of the prepuce and may become a source of irritation and infection.

There is a pocket within the glans above the urethra called the urethral diverticulum. A buildup of putty-like smegma in this pocket, called a bean, can compress the urethra and cause spraying during urination. The problem is treated by manually everting the urethral diverticulum and removing the accumulated smegma.

BREEDING SOUNDNESS EXAMINATION OF THE STALLION

The BSE is a physical examination performed by your veterinarian before the stallion is first used at stud. It is usually repeated prior to each breeding season and whenever there is a question about the stallion's fertility. It is often required as a condition of the stallion's sale. Veterinary evaluation of the reproductive tract is the main focus of the BSE.

HISTORY

The medical history includes notes on past breeding problems and any recent illnesses. A history of injury to the testicles or penis is significant, because such injuries can interfere with fertility and willingness to breed. It is important to ask whether the stallion was ever given anabolic steroids or testosterone, since these drugs can cause testicular degeneration.

Positive and negative experiences during past breedings may influence the stallion's present attitude toward sex. It is also important to be aware of undesirable vices, such as biting and kicking, as these vices will require additional handling precautions.

GENERAL HEALTH AND APPEARANCE

A breeding stallion should posses excellent health. A stallion suffering from chronic illness or a heavy burden of internal parasites is a poor candidate for breeding. Sperm production and sex drive are both diminished by illness.

The well-conditioned stallion should have a body score of about 5 (see BODY-CONDITION SCORE in the appendix). Fat stallions are sluggish. Thin stallions often lack libido and may not produce adequate quantities of sperm.

Vital signs should be taken to rule out anemia and acute infections. Listening to the lungs will determine if there are signs of respiratory disease. The presence of a heart murmur requires a diagnostic workup.

The musculoskeletal examination may reveal a back disorder, wobbler syndrome, crooked legs or other bone and joint problems that could be inherited. Laminitis and diseases affecting the feet, back and hindquarters can make it difficult for a stallion to mount a mare or remain in a mounted position long enough to achieve ejaculation.

A dental examination should be part of the BSE. It not only identifies dental disease but also confirms the approximate age of the stallion.

GENITAL EXAMINATION

Both testicles are examined for normal size, shape and consistency. When one testicle is missing from the scrotum, the stallion is disqualified for breeding because cryptorchidism is heritable; see UNDESCENDED TESTICLES (CRYPTORCHIDISM) in chapter 10. An exception might be made for a stallion who had one testicle removed for medical reasons. Even so, his sperm output will be reduced by 50 percent. Compensatory enlargement of the remaining testicle does not occur in the mature stallion, although it may occur in the very young stallion.

A most important objectives of the BSE is to determine a stallion's potency, or the number of mares he may be expected to breed in a season. The quantity of sperm he is capable of producing is directly related to the size of his testicles. A stallion with small testicles is not as potent as one with large testicles.

Testicular size is related to the width of the scrotum. Total scrotal width (TSW) is measured using a pair of calipers applied across the widest part of the scrotum. The average TSW for light horse breeds is about 4 inches (approximately 100 mm). A mature stallion with large testicles might have a TSW of 5 inches (127 mm). A stallion with small testicles will have a TSW of about 3 inches (80 mm).

The testicles do not begin to develop and produce sperm until a stallion is about 16 months of age (the range is 1 to 2 years). The testicles then continue to grow until age 7. A 2- or 3-year-old stallion will not have the testicular size or reproductive capacity of a 4-year-old stallion.

A second important determinant of potency is the consistency of the testicle as determined by feeling each testicle between the thumb and fingers. The normal testicle is oval, smooth and slightly firm. Testicles that are either too soft or too hard are often associated with testicular degeneration and reduced or absent sperm production. A firm, asymmetrically enlarged testicle is suspicious for tumor.

Occasionally a stallion is found to have a testicle rotated in the scrotum. This does not appear to interfere with sperm production or the quality of his semen (see TORSION OF THE SPERMATIC CORD in chapter 10).

Each epididymis is palpated to determine if it is normal, large, small, hard or soft. The epididymis is the main storage area for sperm. A hard area in the epididymis suggests scarring.

The inguinal canals are examined for hernias.

The penis and prepuce are cleaned and examined for scars, pustules and growths. Cultures are taken with a cotton-tip applicator from the end of the urethra, the urethral diverticulum, and the shaft of the penis. Immediately after ejaculation, the urethra is again cultured, as is the semen.

The prostate and accessory sex glands are examined by rectal palpation.

SEMEN ANALYSIS

Veterinary collection and analysis of semen is the most important part of the reproductive examination. This highly technical aspect of stallion management is best carried out by an equine practitioner.

The initial purpose of the semen analysis is to determine if the stallion is producing enough progressively motile sperm to impregnate a mare. If not, is the condition treatable?

A second purpose involves planning the stallion's book. Given the number and quality of sperm, how many mares can the stallion be expected to cover per day?

These two questions are best answered by analyzing the semen from two ejaculates collected one hour apart, or from a single ejaculate collected daily for 6 or 7 days. The 6-day test is more informative as to the stallion's daily sperm output and the number of mares he can be expected to breed.

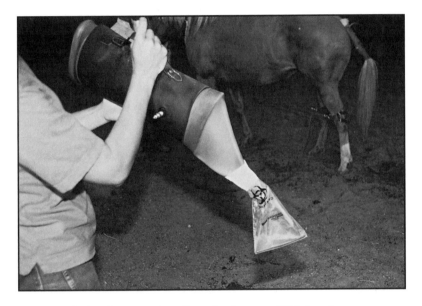

Semen analysis. The semen is collected using an artificial vagina.

The semen is collected using an artificial vagina (AV) by techniques described in chapter 6. Immediately after collection, the semen is transported to the laboratory. All equipment and supplies that may come into contact with the semen are warmed to body temperature.

Ejaculated semen is composed of a gel fraction and a gel-free fraction. The larger, gel-free fraction contains the important seminal fluid and sperm. If the AV did not contain an in-line filter to remove the gel, the gel is removed by passing the semen through a nylon filter. The gel fraction can also be removed by carefully aspirating it with a syringe.

The gel-free semen then is analyzed for total sperm count (expressed in billions), motility of the sperm cells (expressed as a percentage), and percent of sperm that are normal in appearance. The semen is also checked for pH and cultured for bacteria.

It is important to note whether there are contaminants in the seminal fluid, such as bacteria, white blood cells, urine or blood. Urine is recognized by yellow-colored semen, and blood by a pink or reddish-colored semen. Urospermia and hemospermia are discussed in chapter 10.

The volume of gel-free semen is measured using a graduated cylinder. Normal volumes range from 20 mL to greater than 300 mL, with an average of 50 to 60 mL. The amount of fluid ejaculated does not necessarily reflect the total number of sperm. Stallions who ejaculate large volumes of seminal fluid actually have a lower concentration of sperm per mL than do those who ejaculate small volumes of semen.

A sample of semen is placed in an instrument called a densimeter, which measures the concentration of sperm in 1 mL of semen. The total number of spermatozoa in the ejaculate is obtained by multiplying the concentration per mL by the volume of the semen. A typical fertile stallion might produce 10 to 15 billion sperm per ejaculate. The number of spermatozoa in a second ejaculation 1 hour later is approximately 50 percent of that present in the first.

Shortly after ejaculation, the semen should be examined for percent motility. The first step is to dilute the semen with a prewarmed commercial or laboratory-prepared semen extender. Sperm cells in raw, undiluted semen tend to clump, making it difficult to estimate the percentage of progressively motile cells.

The effect of temperature on the motility of sperm is pronounced. The highest motility is noted when the semen is examined within minutes of ejaculation while still close to body temperature (37°C). If the semen is allowed to cool to room temperature, as much as 30 percent of potentially motile sperm may be excluded from the calculation. Accordingly, if the specimen cannot be examined immediately, it should be put into an incubator at 37°C. However, incubation of extended semen for several hours at this temperature is counterproductive since it greatly decreases motility and shortens the life of the sperm.

The two methods of estimating percent motility are human observation and computer-assisted analysis. Human observation involves placing a drop of semen on a slide, covering it with a slip, and estimating the percentage of progressively motile sperm as viewed in a number of fields. A phase-contrast microscope with a heated stage is best for this purpose.

Many sperm cells that are moving may not be progressively motile. Sperm swim with different velocities and often pause to rest. Stallion sperm frequently swim in small circles but make no progress. Others may swim backward. To be counted as progressively motile (the accepted standard), a sperm must be moving rapidly across the microscopic field and making forward progress with each lash of its tail. The semen of a fertile stallion might contain 60 to 85 percent progressively motile sperm.

The alternative to human observation is to use an electronic counter with a computer-assisted program that recognizes the various motility patterns and gives an estimate of the percentage of sperm that is progressively motile. The electronic counter eliminates the subjective factor associated with human observation. However, with both methods, only a rough estimate of the percentage of viable sperm is possible.

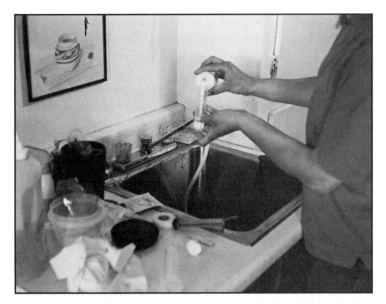

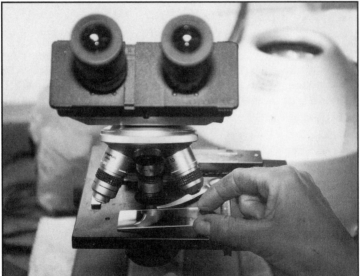

The specimen is diluted with semen extender and then examined under the microscope to determine percent motility.

Morphology refers to the external appearance of the sperm. A sperm cell is composed of an oval-shaped head, a midpiece and a long, straight tail. Morphologically abnormal sperm are sperm without heads or tails; sperm that have double heads; double tails; and tails that are bent, coiled or small. A fertile stallion generally will have at least 60 percent morphologically normal sperm.

Attempts have been made to compare the percent of abnormal sperm in the ejaculate with the fertility of the stallion. Exact guidelines have not been established. There is disagreement among observers as to what constitutes an abnormal sperm and also lack of knowledge regarding the structural defects that interfere with fertilization. However, if a high percentage of sperm appear deformed, the stallion's fertility may well be low.

Cultures of the urethra and semen frequently reveal the presence of bacteria. Most bacteria found in the stallion's urethra and semen do not interfere with the quality of the semen or the stallion's fertility. For more information, see SEMEN CAUSES OF INFERTILITY, chapter 10.

MANAGING THE STALLION FOR MAXIMUM REPRODUCTIVE EFFICIENCY

A valuable stallion with a large booking must be managed efficiently to maximize the number of mares he can impregnate during the breeding season. The question is, how often can he be used at stud without compromising his potency or sex drive?

The two most important aspects of a stallion's fertility are the total number of sperm he can produce in each ejaculate and the percentage that is progressively motile.

As noted earlier, sperm production is related to the size of the stallion's testicles. It is also greatly influenced by season. Early in the breeding season, the daily sperm output may be only 50 percent of what it will be at the peak of the season.

Increasing the photoperiod by using an artificial light program, as described for the mare in chapter 1, has been shown to increase the daily sperm output of stallions for the months of February through June. To be sensitive to photostimulation, the stallion must be allowed to experience the normal decreasing day lengths of the preceding fall. Having the stallion's sperm production peak in early spring, however, might not be advantageous if most mares are going to be bred in late spring and summer.

A study designed to examine the number of sperm in successive ejaculates showed that the volume of semen as well as the number of spermatozoa per ejaculate increase with the age of the stallion. Stallions who were 2 and 3 years old produced fewer sperm than did stallions 9 to 16 years of age. However, the quality of the semen as evidenced by percent motility did not differ between the two groups. Young stallions did not, in fact, have fewer motile sperm, nor did they produce semen of inferior quality.

One conclusion of the study was that stallions do not achieve maximum reproductive potential until at least 6 years of age. Once they attain this potential, they appear to maintain the same potency for many years.

Frequency of ejaculation has long been known to be one of the most important factors affecting daily sperm output. With frequent ejaculation (once or more per day), the number of spermatozoa in each ejaculate begins to fall and may eventually reach numbers below those considered critical for fertilization (200 million sperm per ejaculate). However, when ejaculation occurs no more frequently than every other day, the total number of sperm does not fall with each ejaculate.

The capacity of an individual stallion to maintain adequate sperm numbers with frequent usage can be determined by collecting his semen daily for 14 days. During the first 7 days, the results will be affected by withdrawals from the stallion's sperm reserves stored in the epididymis. After 7 days, these reserves are depleted and the daily sperm output corresponds to his capacity to produce sperm.

As an example, a fertile stallion who produces 13 billion sperm on the first day may produce only 2 or 3 billion sperm on the 7th through the 14th day. These later numbers should not preclude frequent usage by live cover during the spring and summer. However, note that if this same stallion were collected every other day, his ejaculate would always contain 13 billion sperm, or at least twice the number that could be obtained by collecting him daily. This can be an important consideration when trying to maximize the number of sperm per ejaculate.

For example, if a stallion is being used in an artificial insemination program employing fresh semen, an efficient collection frequency might be every other day. Part of the semen is used to inseminate mares who have been in standing heat for at least 2 days. The remainder is used to reinseminate mares who are still in heat.

If the semen is frozen and stored, maximizing the number of sperm per ejaculate is even more important. To take advantage of semen reserves, a collection frequency of twice a week might provide the maximum number of sperm per ejaculate.

In summary, the actual number of mares who can be bred by natural service or artificial insemination varies with the individual stallion. With a mature, fertile stallion, the daily sperm output is usually adequate to allow him to cover two or three mares a day. For the young stallion, two or three times a week would be a reasonable frequency. In actuality, it is the stallion's sex drive, rather than his sperm-producing capacity, that often is the principal factor in limiting the number of mares he can breed by natural service. A stallion should be carefully managed to maintain his sex drive throughout the breeding season. This requires periodic evaluation of his breeding behavior and using him less frequently or giving him time off if his libido begins to drop.

CARE OF THE STALLION

FEEDING AND NUTRITION

The dietary needs of an individual stallion are determined by his weight, body condition, activity level and temperament. In general, except for an increase in dietary energy, the nutritional requirements of a stallion during the breeding season do not differ from those shown for the mature horse at maintenance (see Table I, appendix).

A recent report from the National Research Council states that a breeding stallion requires 25 percent more digestible energy (DE) per day than he requires for maintenance. This would be the equivalent of adding 5 pounds of hay or 3$^1/_3$ pounds of grain to his daily ration. Depending on his activity level, this additional dietary energy may be too little or too much. In actual practice, stallions are frequently overfed and oversupplemented, despite the fact that excess dietary energy is associated with obesity, laminitis, and lack of libido. To ensure optimal weight and a body-condition score of 5 (see BODY-CONDITION SCORE, appendix), the stallion should be weighed at least twice a month and his energy needs adjusted accordingly.

Contrary to popular belief, a stallion's protein requirements do not increase during the breeding season. However, it is important to provide high-quality protein by using only the best feeds.

The diet should contain adequate amounts of vitamin A for healthy sperm production. Good forage supplies ample amounts of this vitamin. However, a deficiency could occur if a stallion was not able to consume green hay or forage for several months. This would be an indication for giving a vitamin A supplement.

Vitamin E has been credited with improving sperm function and libido, but there is no scientific evidence to support these claims.

Access to clean, fresh water and trace-mineralized salt should be provided at all times.

VACCINATIONS

Because a stallion is exposed to many outside mares, it is essential that all his vaccinations be up-to-date before the breeding season begins. A recommended immunization schedule is shown in Table II, appendix. Boosters for equine influenza and rhinopneumonitis should be given every 3 or 6 months, depending on the prevalence of disease in the area.

Maintain protection against Eastern, Western, and Venezuelan encephalomyelitis. Annual boosters are required, preferably given 1 month before the height of the mosquito season. An annual rabies vaccine is recommended in rabies-endemic areas. Additional vaccinations recommended for horses in high-risk areas are shown in Table III, appendix.

DEWORMING

Internal parasites cause weight loss, anemia, diarrhea and reduced fertility. The stallion should be on a routine deworming program, as described in INTESTINAL PARASITES in chapter 7. No additional deworming is required in preparation for breeding. All currently recommended anthelmintics can be safely given to stallions without fear of harming sperm production.

EXERCISE

A stallion needs daily exercise for cardiovascular fitness and breeding stamina. Regular exercise maintains weight and muscle tone, prevents boredom and appears to improve libido.

It is not necessary for a breeding stallion to be in top athletic condition. In fact, if a stallion has been involved in intensive training, a period of physical letdown is desirable.

PREVENTIVE DENTISTRY

Young to middle-age horses develop sharp points on their molars that, if not kept in check by periodic filing, cause mouth pain and can lead to abnormal chewing patterns. This can result in failure to extract the full nutritional value from the meal. A program of good oral hygiene, as described in CARE OF THE BROODMARE in chapter 1, will prevent these problems.

FOOT CARE

Care of the feet is especially important. Pain and lameness make it difficult for a stallion to mount or remain in a mounted position long enough to achieve ejaculation. The feet should be inspected and cleaned every day, and the hooves trimmed every 4 to 6 weeks. Remove the stallion's front shoes to prevent injury to the mare.

HANDLING THE STALLION

Stallion handling requires self-confidence, an understanding of stallion psychology, the ability to anticipate what a stallion is likely to do, and the knowledge of how and when to apply physical restraint. These skills are learned through experience.

A handler who lacks confidence in his or her ability to remain poised and in command of the situation may feel compelled to employ excessive force or unnecessary punishment to maintain control—actions that only confuse or frighten the stallion and ultimately make him more difficult to handle. A cycle develops in which the stallion becomes ever more intractable and thus the recipient of even greater abuse.

Excessive, inconsistent, and unnecessary punishment causes many stallions to resent handling, resist restraint, and fail to display normal mating behavior. In fact, handling errors are among the most important causes of loss of stallion libido. For more information, see STALLION HANDLING MISTAKES in chapter 10.

A well-trained stallion, socialized early in life, knows that obedience and submission are the normal conditions between horse and master. When corrective discipline is required, it should be dispatched promptly and without anger, after which the stallion should be required to perform a familiar task, such as backing or turning. This reestablishes the authority of the handler.

The essential tools for discipline and control are the lead shank, chain and halter. The shank connects to a 2-foot chain with a snap at the end. The snap attaches to the halter. All parts must be stout, in good repair, and should be replaced when worn. The lead can be made of rope, nylon or leather. Leather is less likely to cause friction burns to the hands.

A chain placed across the gums exerts pressure on an extremely sensitive area.

A chain over the nose or through the mouth is a moderately severe restraint.

Control is determined by how the chain is placed. A chain placed across the gums of the upper teeth exerts pressure on an extremely sensitive area and will cause considerable pain. A chain placed over the nose or through the mouth and jerked down hard also exerts substantial pain and gains the stallion's immediate attention. However, chains so placed may cut the gums, nose or corners of the mouth. A chain placed under the chin is less severe and suffices for many situations.

Some handlers prefer to use a Chifney bit instead of a chain. A Chifney bit is a loose round ring placed in the mouth so that it encircles the lower jaw. The bit is attached to the halter. The lead rope connects to the bit beneath the chin. The effect of pulling on the lead is similar to that of pulling on a chain through the mouth.

The placement of the chain or bit is one part of the equation. The other is how forcefully the pressure is applied. An experienced handler will use the least amount of force necessary to control the stallion.

A chain should be removed whenever a horse is tied. If the horse becomes frightened and pulls back, the chain will constrict and cause a serious injury.

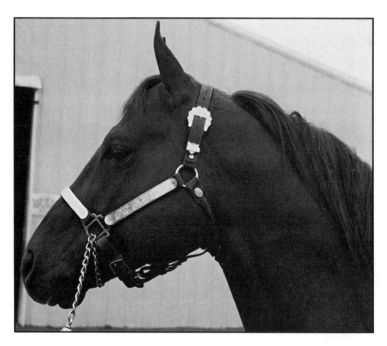

A chain beneath the chin is a milder restraint and suffices for many stallions.

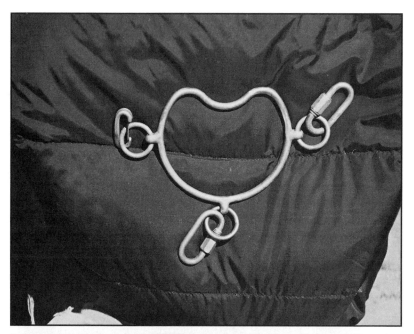

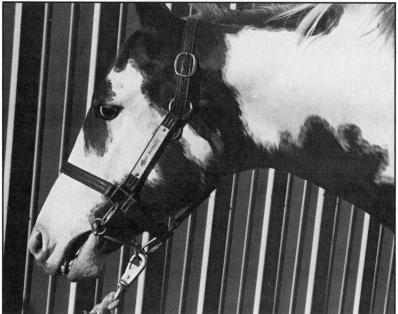

A Chifney bit. The effect is like that of a chain through the mouth, with less likelihood of cutting the lips.

STALLION VICES

Many stallions have vices. It is important to have a knowledge of dangerous stallion behavior and a plan for dealing with such behavior. Even a dependable stallion is capable of doing something totally unexpected when in a state of sexual arousal.

Misbehavior often evokes anger and the desire to punish. Resist the impulse to discipline sharply unless you have the physical restraints in place to guarantee that you will win in a nose-to-nose confrontation. Once a stallion realizes that he is physically stronger and can get away with what he wants, he becomes extremely difficult to manage.

Many aggressive vices arise out of natural male behavior and can be anticipated in the context in which they occur. This type of misbehavior is not viciousness. The immediate objectives are to establish control and prevent injury. Corrective punishment may reduce the frequency and severity of the misbehavior.

Occasionally one encounters a horse of truly vicious temperament. This individual is unpredictable, dangerous, and must be handled at all times with extreme caution. This horse is not suitable as a companion or pleasure horse and is a bad candidate for breeding. If the horse is a stallion, gelding can be considered. However, there is a chance that gelding will have no effect on his behavior.

BITING

Biting is the most common vice of horses. A stallion may bite his handler, bite the mare, or bite another horse. A horse getting ready to bite often lays back his ears and raises his upper lip. However, many horses bite without warning.

All attempts at biting should be met with a sharp downward jerk on the lead shank and/or a forceful rap on the muzzle accompanied by a harsh word. Do not swat at the horse's face. Slapping is inaccurate and can make the horse head-shy. Corrective punishment must be administered as soon as the horse starts to bite. All personnel who handle the horse should be aware of this vice and be prepared to administer discipline.

Biting and nipping are much easier to prevent than to cure. The explorative, nippy behavior of a young colt or filly can turn into a bad habit if allowed to go unchecked. Activities that foster nipping and biting should be discouraged. Do not feed treats by hand or allow a foal to nuzzle and lip its handlers. Avoid carrying apples and carrots in your pockets.

Never assume that a biter has been cured. This vice is hard to correct.

REARING AND STRIKING

Rearing and striking are potentially lethal. Situations in which rearing may occur must be anticipated so that measures can be taken to protect people and horses.

Stallions fight for herd dominance by rearing on the back legs and striking with the front feet. A stallion who has been allowed to dominate his keepers may rear to avoid being caught. Rearing can also occur when a stallion is cornered and frightened. Excited stallions frequently rear and paw when introduced to a mare in heat.

Many horses give warning by tensing up; others do not. Be aware of situations in which rearing may occur. A horse is most likely to strike at another horse when in nose-to-nose confrontation. Always stand well to the side and never at any time between the two horses.

A stallion with a history of rearing should be controlled by passing the chain over the nose or between the lips, as described above. Do not place the chain under the chin as this can actually promote rearing.

To control a rearing horse, jerk down strongly on the lead shank (either as the horse begins to rear or when he starts to come down), but not while he is standing on his hind legs. If he resists and throws his head back, he may lose his balance and fall over backward. After pulling down hard on the shank, maintain tension on the lead and quickly back the stallion away. This reestablishes control and prevents further attempts at rearing.

Kicks and punches to the stallion's belly have been cited as punishment for rearing. Only very agile and athletic people can administer an attention-getting blow to the belly of a rearing stallion. Kicks, in particular, usually cannot be delivered with much force and may leave the kicker off-balance and vulnerable.

KICKING

Kicking is another serious problem that can cause injury and death.

Kicking frequently occurs in situations that follow a pattern and can be anticipated. For example, a horse may kick when disturbed during a meal, during saddling as the cinch is tightened, when handled about the hindquarters, or when startled from the rear. The handler who anticipates the kick can be ready to deliver punishment in the form of a jab to the ribs or a jerk on the lead shank accompanied by a loud reprimand.

Stallions may kick other horses and people for the same reasons they rear and strike.

A stallion used as a teaser may kick at a mare out of sexual frustration. Some stallions who have been kicked by mares during teasing or breeding may kick preemptively to establish dominance.

Be prepared for either horse to kick during live cover. The mare is most likely to kick as the stallion dismounts. *Caution: It is imperative that you do not stand behind a mare for any reason at any time during live cover.*

CHARGING

Charging or lunging is when a horse attacks or attempts to savage an attendant in a stall or paddock. This type of deliberate viciousness has many causes, but improper human socialization and rough treatment appear to be at the root of most cases. Only the most experienced stallion manager should attempt to break a horse of a deeply ingrained aggressive vice such as charging.

MASTURBATION

A stallion masturbates by thumping his erect penis against his belly. For many years, it was assumed that masturbation depleted sexual energy and sperm reserves. Recent studies show that masturbation is a normal sexual activity for stallions and that it does not deplete sperm reserves or interfere with libido. In fact, ejaculation rarely takes place. Also contrary to popular belief, masturbation does not appear to be related to, or influenced by, frustration or boredom.

Stallion rings, brushes and/or cages have been used to prevent erection and masturbation and to discourage stallions from displaying sexual interest in mares. The stallion ring fits snugly around the penis just behind the glans. Problems associated with rings and brushes include penile injury and infertility related to blood in the semen. Use of these devices may lead to abnormal sexual behavior and breeding problems. For these reasons, they are not recommended.

STABLE VICES

Cribbing, wood-chewing, stall-kicking, head-nodding, pawing and weaving are stereotypic activities that occur primarily in stabled horses. A stereotypic behavior is a modification of a normal activity in which the pattern is repeated constantly and the behavior appears to serve no purpose. Stereotypic behaviors do not occur in wild horses. They are believed to arise out of boredom, along with close confinement and lack of species interaction—conditions known to produce frustration and discontent. In a few cases, the horse spends so much time in the stereotypic activity that he fails to consume enough feed and loses weight and condition. A hereditary predisposition for cribbing, weaving and stall-circling has been identified in some Thoroughbred families.

Cribbing is an activity in which a horse sets his front teeth on a ledge, arches his neck and then pulls back and swallows large amounts of air. Severe cribbers wear down their incisors and eventually may be unable to graze.

The usual treatment for cribbing is to apply a 2- to 3-inch strap around the horse's throat at the narrowest part of the neck. As the horse arches his neck, the strap cuts into the throat and causes pain. The strap is fitted snugly but not so tight as to interfere with breathing. Crib straps are often effective. However, some horses continue to crib in spite of the strap, and most horses will start cribbing again as soon as the strap is removed. Various surgeries have been developed to prevent cribbing. Results are not always successful.

Wood-chewing is similar to cribbing, except that when the horse sets his teeth, instead of swallowing air, he bites off pieces of wood. Wood-chewing usually does not cause impaction or other intestinal problems, because most of the wood is dropped and not swallowed. Splinters can penetrate the soft tissues of the mouth and cause infection and abscess. Wood-chewing is tremendously destructive of fences, stalls and wooden siding. A few cases of wood-chewing may be caused by lack of adequate roughage in the diet.

Wood-chewing is difficult to control. These horses, even when turned out to pasture, often continue to destroy fences and posts. Electric wire can be strung along the top of wooden fences. Posts can be painted with a nontoxic chemical—not creosote—to make them distasteful. Creosote contains substances that can cause poisoning. If the horse is eating pelleted feed, switch to hay, or feed a pelleted ration along with one-half pound of long-stem hay per 100 lb. weight of the horse.

Horses who chew wood and crib should be separated from other horses who may imitate the behavior and acquire the habit.

Stall-kicking is when a horse repeatedly kicks at the wall of his stall without apparent cause. It should be distinguished from wall-kicking, which is caused by the presence of a horse in the next stall whom the kicker resents. Stall-kicking often results in chronic stress injuries to the horse's feet and limbs.

Pawing is a behavior associated with frustration. A stallion prevented from mounting a mare will often paw the ground. The same is true for a horse who wants to move forward but is held back.

Weaving is when the horse stands in one place and constantly swings his head from side to side while shifting his weight from foot to foot.

Stall-circling is when a horse continuously paces around the inside of his stall or paddock, sometimes in a figure of eight.

Tail-rubbing is a form of self-mutilation in which the horse rubs up against a wall or post. Hair is lost from the base of the tail, and the skin becomes raw and excoriated. Tail-rubbing can be caused by hypersensitivity to the bites of gnats and flies, and by the presence of pinworms and tail mites. Medical causes should be excluded before concluding that the tail-rubbing is behavioral.

Head-bobbing and head-shaking are activities that normally serve to dislodge flies. A horse who repeatedly raises and lowers his head or shakes it from side to side when flies are absent is engaging in stereotypic behavior—assuming there is no medical cause. Head-bobbing can be associated with encephalitis as well as diseases of the middle and inner ear. Head-shaking is seen with external ear infections and foreign bodies such as ticks in the ear canal.

Stereotypic behavior is easier to prevent than treat. Once a vice has become firmly established, the behavior often persists even after the cause has been removed.

When possible, horses should be turned out to pasture with other horses. Exercising the horse on a walker followed by an hour spent in a dirt paddock provides good exercise and alleviates boredom. Commercial or homemade horse toys help to distract horses confined to stalls. Hanging a plastic jug filled with pebbles gives the horse something to play with. Putting up mirrors makes the horse feel as though he has a companion.

GELDING THE STALLION

Gelding involves the complete removal of both testicles. The purpose of the operation is to prevent or eliminate masculine behavior in a horse not intended for breeding. A gelding is more tractable than a stallion and not as easily distracted by other horses during events. He is also much easier to keep with other horses and shows little or no interest in estrus mares.

A horse can be gelded at any age. Most males are gelded at 12 to 18 months of age, when they have achieved 90 percent of their adult stature.

Veterinary examination should be done before a horse is gelded. The horse who has not been on a parasite-control program should be dewormed. All horses are given a tetanus toxoid booster. Horses without prior tetanus immunization should also be given tetanus antitoxin.

THE OPERATION

Surgery is done under IV sedation and local anesthesia, with the horse standing in stocks or lying on his side. An incision is made in the scrotum. After removing both testicles, the scrotal incision is left open to heal from within. Postoperative complications are infrequent. They include bleeding from the testicular artery, scrotal *hematoma*, spermatic cord infection, abscess of the scrotum and rupture of the small intestine through the scrotal incision.

Postoperatively the horse should be confined to a clean stall or paddock and protected from flies. It is important to walk the horse 15 to 30 minutes three times a day. This helps to prevent scrotal swelling and infection. Continue the exercise until the scrotum is healed.

Notify your veterinarian if the horse develops colic, fever, or increasing swelling or drainage at the site of the operation.

When a stallion is castrated after 3 years of age, his libido and sex drive usually persist for many months. Because sperm remain in the reproductive tract for a variable time, geldings are considered capable of impregnating mares for up to 60 days. As a precaution, separate a gelding from mares for the first 2 months after surgery.

UNILATERAL GELDING

The indications for removing one testicle include testicular injury, torsion of the spermatic cord, varicocele, testicular tumor, inguinal/scrotal hernias, and orchitis. These subjects are discussed in "Stallion Infertility," chapter 10.

If the opposite testicle is normal and healthy, the stallion will continue to produce sperm and have normal testosterone levels. When a testicle is removed before sexual maturity, the remaining testicle often undergoes compensatory enlargement.

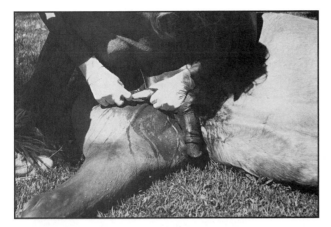

Gelding the stallion. Under IV sedation and local anesthesia, an incision is made in the skin of the scrotum and the testicle is delivered.

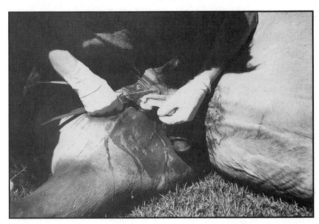

The cord is crushed with an emasculator clamp to prevent bleeding. The cord is then divided beyond the clamp.

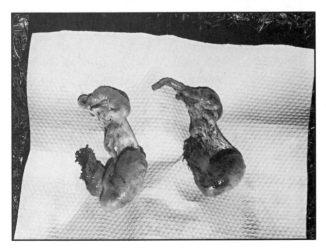

The testicles are examined. Note in each case the intact epididymis (above) and the testicle below. All functioning testicular tissue has been removed.

To prevent accidental pregnancy, separate the gelding from mares for the first 2 months after surgery.

PERSISTENT STALLION BEHAVIOR (PROUD CUTS)

There is a syndrome that follows gelding in which mature stallions (and even males gelded at 1 to 2 years of age) exhibit masculine behavior that persists indefinitely. This behavior includes male aggression, mounting mares, penetrating and even ejaculating. These geldings are referred to as proud cuts, suggesting that failure to remove both testicles is the cause of the male behavior. While incomplete gelding does account for some of these cases, in many cases the horse was properly gleded and there is no explanation for the persistent stallion behavior.

One error that can occur during gelding is not finding a testicle on one side and assuming that it never developed. In fact, a testicle was present but not in its usual location; see UNDESCENDED TESTICLES (CRYPTORCHIDISM), chapter 10. Other infrequent errors include mistaking an epididymis for a testicle and removing only the epididymis, and leaving the epididymis with a remnant of testicle behind.

Two diagnostic blood tests can be used to determine if testicular tissue is present. One test involves drawing serum testosterone levels before and after injecting the horse with human chorionic gonadotropin (HCG). HCG causes the testicles to manufacture and release testosterone. The test is positive when there is a two- to three-fold increase in the concentration of testosterone after HCG. This test is most useful in males 2 years of age and younger. In horses 3 years and older, an elevated serum estrone sulfate concentration alone is diagnostic of retained testicular tissue.

If blood tests are positive, an ultrasound may reveal the location of the retained testicle. This information helps in planning where to make the incision to find and remove the remaining testicle or its remnant.

Chapter 4

DETERMINING
WHEN TO BREED

THE IMPORTANCE OF KEEPING GOOD RECORDS

It is important to keep a complete set of records on every horse. Mare records are particularly important because the more that is known about each mare's breeding history, the easier it will be to breed her.

Each mare has her own individual estrous behavior. For example, some mares ovulate on day 4, others on day 7. This pattern tends to remain relatively constant and predictable throughout the mare's reproductive lifetime. Furthermore, mares tend to ovulate the same-size follicle year after year. Breeding records that contain a mare's approximate ovulatory follicle size and ovulation pattern are extremely useful in predicting the best time to breed. Similarly, the number of days in estrus, the interval between estrous periods, and the interval between successive ovulations are extremely important data and should be readily accessible to the farm manager and veterinarian.

Mares exhibit wide variation in their responses to teasing. Some mares display all the signs of estrus several days before they ovulate. Other mares show only subtle signs, or appear to be in heat for only one or two days. The period of receptivity could be missed without a good teasing history.

A comprehensive set of records should provide identification data and include the horse's entire healthcare from birth to the present. This information should be

---

MARE HEALTH RECORD CHECKLIST

Registration Data. The horse's name; breed; date of birth; sire and dam; breed registration number; and owner's name, address and telephone number.

Visual Identification. Color (bay, sorrel, Palomino), size (height), brand, tattoo, identifiable scars or physical characteristics.

Health Information. Vaccinations, deworming, dental records, foot care, results of Coggins and other tests, adverse effects of past illnesses or injuries, any known allergies, hormone administration at any time in the past.

Breeding and Reproductive History. Results of past breedings (non-conceptions, abortions, live foals), length of heat cycles, number of days in estrus, interval between ovulations, approximate ovulatory follicle size, history of abnormal heat cycle, infertility problems (abortion, endometritis, biopsies, culture), breeding behavior problems, Caslick's operation (open or closed).

---

available to the farm manager, who will be responsible for assuming continuity of care during the mare's temporary residency.

If there is a history of infertility, the health information should include the results of diagnostic studies (especially endometrial biopsies and cultures), before and after treatment.

It is essential to keep a record of the mare's teasing responses (see HEAT DETECTION AND TEASING, below) and findings on rectal palpation. These records can be extremely helpful in predicting when breeding is most likely to result in pregnancy.

Rectal palpation focuses on the ovaries, cervix and uterus—particularly on the size and consistency of dominant follicles. A dominant follicle will approach ovulation size at about 40 mm. The consistency of the follicle (whether it is turgid, soft or very soft) helps to predict how soon the follicle will ovulate.

A foal's record should be maintained for every mare with a foal at her side. This record should be part of, or appended to, the mare's record. The purpose of this record is for foal identification and to maintain a database for vaccinations and dewormings given during the foal's residence. If the foal was born on the premises, it is a good idea to take a picture and send a copy to the mare's owner. This aids in identification and also is good public relations.

BREEDING RECORD

Important Dates. Day arrived, day in heat, day bred, pregnancy exam, pregnancy confirmed, treatments given, day departed.

Teasing Data. Daily tease score or similar notation (e.g., in heat, not in heat). Can be usefully combined with palpation data in chart form.

Palpation Data. Date-specific chart showing results of palpation of right and left ovaries (follicle size and consistency, ovulation depression, corpus hemorrhagicum), uterine tone (excellent, good, fair, poor), and cervix (percent relaxation).

FOAL'S RECORD

- Sex, color, markings
- Birth date and time
- Vaccination/dates
- Deworming/dates
- Photograph

HEAT DETECTION AND TEASING

Heat is the period of estrus. It lasts an average of 5 to 7 days. It is characterized by increasing sexual interest and receptivity to breeding. During estrus, one or more dominant ovarian follicles develop and secrete estrogen, the hormone responsible for the signs of sexual receptivity. When the follicle collapses in the act of ovulation, the blood estrogen level falls, and within 24 to 48 hours the mare goes out of heat. At this point she becomes unfriendly to the stallion. Thus ovulation is said to mark the end of the heat phase of the estrous cycle.

Ovulation is a specific event that can be detected by rectal palpation, ultrasonography and measurement of serum progesterone. After ovulation, the follicle converts to a corpus luteum (CL) and begins to manufacture progesterone, the hormone that prepares the uterus to accept and nourish the embryo. Thus elevated serum levels during the natural breeding season of progesterone are found in the diestrus phase of the estrous cycle.

If a mare does not show interest in the stallion during the natural breeding season, she is quite likely exhibiting the common condition discussed below known as silent heat. Silent heat is a normal heat cycle that is not detected. Mares who do not show heat merit veterinary investigation (see chapter 2, "Abnormal Estrous Cycles").

SIGNS OF HEAT

The easiest way to determine whether a mare is in heat is to expose her to a stallion. The behavior of the mare communicates her readiness to breed.

Most mares in heat exhibit one or more characteristic signs that include urination, winking, tail raising, squatting, willingness to stand, and friendly behavior toward the stallion. Winking is turning out of the vulvar lips with presentation of the clitoris. Squatting is stretching out the back legs and lowering the pelvis. Willingness to stand is when the mare presents her hindquarters, raises her tail and stands fast while the stallion mounts. Friendly behavior is when the mare pricks her ears forward, remains calm, and stands quietly and nuzzles the stallion.

Note that not all mares display estrus in a typical fashion. Each mare has her own individual pattern. The detection of estrus is not a science. It is an art that improves with experience.

In contrast, a mare who has gone out of heat demonstrates unfriendly behavior such as pinning her ears back, moving about restlessly, switching her tail, and often kicking, striking and biting at the stud. Stallion rejection is diestrus behavior. It appears during the first 2 days after ovulation. By the 3rd or 4th day, nearly all mares will vigorously reject the stallion.

The mare's response to teasing can be coded as a tease score, with numbers assigned to degrees of receptive behavior. In most codes, a score of 1 is given if the mare aggressively rejects the stallion. If she remains indifferent, she is given a score of 2. If she seems somewhat interested, she rates a score of 3. A score of 4 indicates that she is responsive to the stallion and exhibits winking of the vulva and frequent urination. If she urinates profusely, strikes a breeding stance and presents her hindquarters, she is given a score of 5.

The likelihood that a mare will exhibit heat increases as ovulation approaches. Among all mares who exhibit heat, roughly 50 percent will do so by 5 days before ovulation, 75 percent will do so at least 3 days before ovulation, and 90 percent will exhibit heat at least 48 hours before ovulation. Once in heat, the intensity of each mare's estrus response does not appear to change or become stronger as ovulation approaches. Thus there is no reliable way to predict on the basis of the mare's behavior precisely when she will ovulate.

Mares who exhibit neither a positive nor a negative response to teasing can be termed indifferent or passive. A tractable mare with indifferent behavior may allow the stud to mount but will not raise her tail. A passive response generally occurs when the mare is about to come into or go out of estrus.

An estrus mare, showing winking of the clitoris.

This mare rates a tease score of 5.

The combination of positive estrus behavior along with the absence of diestrus behavior is the best indicator of heat. For example, according to one study, a mare who winks, raises her tail, assumes a breeding stance and fails to kick has about an 80 percent chance of being in heat. The other 20 percent of mares are false positives; they show the same type of behavior but actually are not in heat. Thus teasing is only 80 percent specific for identifying heat in the mare.

It is also estimated that up to 15 percent of mares, even on well-managed farms, do not respond to teasing even though they are in heat. Some of these mares may exhibit frank unfriendly behavior such as striking and kicking at the stallion. These mares are difficult to breed and require special veterinary management. See Silent Heat (Behavioral Anestrus) later in this chapter.

The postpartum mare is often less receptive to teasing because her primary focus is on the protection of her foal. This can lead to unfriendly behavior toward the stallion throughout most of estrus. Shy and maiden mares are often reluctant to show signs of heat when teased by an aggressive stallion but may do so when approached by a gentle stallion. In addition, there are mares who do not show heat because of endocrine or behavioral reasons (see Problems with Detecting Heat later in this chapter).

Taking all of the above into consideration, teasing as a reliable indicator of heat has a sensitivity of only about 85 percent. It is a good screening test, but when a higher degree of accuracy is sought, mares should be examined by rectal palpation and/or ultrasonography.

TEASING METHODS

The method you choose will depend on the facilities available, size of the breeding operation and number of personnel available to handle the horses. The primary consideration should always be the safety of the handlers and horses. A good teasing program strives to identify all mares in estrus.

Mares with foals require special attention. Mother and foal should not be separated during the teasing process. Most lactating mares will not show heat if they cannot see and hear their foals. If the mare is going to be teased by the stud without a barrier between them, have a third party hold the foal out of harm's way but close enough to the teasing area to reassure the mother. When one of the horses is in a stall or a secure pen, it is safe to leave the foal at the mother's side.

On many farms, the breeding stallion is used as the teaser. Many stallions enjoy teasing. However, excessive teasing can make some studs aggressive and difficult to handle in the breeding shed. Accordingly, many breeders use a nonbreeding stallion as the teaser. In addition, the use of another male eliminates the risk of injury to a valuable stud. Some breeders use a pony stallion. Pony stallions are easier to manage than horse stallions but are not as readily available.

The temperament of the teaser stallion is an important consideration. He should be easy to handle but aggressive enough to elicit the signs of heat. The ideal teaser

Teasing across a fence. Note the raised tail and receptive behavior of the mare.

courts his mares persistently and does not lose interest halfway through the procedure. The teaser should be controlled with the same restraints as described for handling the stallion (see chapter 3).

A common method of hand teasing involves trying a mare with a stallion across a heavy wooden barrier or metal fence. The barrier should be low enough to allow the stallion to get his head and neck over, but high enough to prevent him from rearing and trying to jump it. A well-padded barrier protects the horses if they kick or strike.

The stallion is led up to the mare, who stands on the other side of the barrier. The horses are introduced nose-to-nose. If the mare is not receptive, she will lay back her ears and swish her tail violently from side to side. It is most important to firmly control the mare and anticipate that she may strike out or swing around and kick at the barrier.

If the mare is receptive, she will lean toward the stallion, spread her back legs, urinate and show winking of the labia.

The mare can also be teased in a box stall with sturdy Dutch doors. Opening the upper door allows the two horses to touch and nuzzle.

Another method of teasing involves leading the stallion up to a paddock containing a group of mares, or leading him down an alley of pens, each containing a single mare.

With the paddock approach, a large number of mares can be teased at one time. It may be less accurate in detecting heat since it depends on each estrus mare voluntarily approaching the stallion. This does not always happen. In fact, aggressive mares often drive off less-assertive mares, leading to suppression of estrus signs in these individuals.

An efficient method of screening for estrus is to lead the stallion down a row of pens, each containing one mare.

This mare is ready and willing to breed.

The kicking behavior of this mare is typical of diestrus.

Pen teasing is much more time-consuming than teasing mares in a paddock but has the advantage that the stallion can stop and investigate each mare individually, allowing the farm manager to note each one's readiness. Occasionally a mare down the line will show estrus even before the stallion gets to her pen. A shy mare may show estrus only after the stallion moves on.

On most breeding farms, mares are teased once a day. If they are teased less frequently, there is a chance of missing the first and last days of estrus. It is important to know both dates in order to breed on the day when fertility is highest. Note also that up to 10 percent of mares do not show heat until 48 hours before they ovulate (the best time to breed). Accordingly, on many breeding farms, as soon as a mare shows estrus, she is examined by rectal palpation and/or ultrasonography to accurately determine when to breed her.

After the mare goes out of heat, it is important to determine whether she is in foal. A successful and early method of detecting pregnancy is to observe the mare's response to teasing 17 to 21 days after breeding. Most early pregnant mares will not show signs of heat. However, if the mare does tease back in, it can be assumed that she is not pregnant. In that case she can be bred back to the stallion immediately without skipping an estrous cycle.

PREDICTING OVULATION

Fertilization is most likely to occur when breeding takes place within the 48 hours preceding ovulation. The average length of survival for fresh sperm within the oviducts is 2 days—although potent sperm have been known to survive for 5 to 6 days. Thus breeding more than 48 hours before ovulation may not produce viable sperm in the oviduct at the time of ovulation. Similarly, breeding more than 12 hours after ovulation is associated with a rapid decline in conceptions. The unfertilized egg loses vitality in a short time. If the egg does not encounter a viable sperm within 8 to 12 hours, conception may not occur. After 12 hours, pregnancy is extremely unlikely.

Breeders who rely on teasing to determine when to breed generally recommend covering the mare on the 3rd day of estrus and every other day thereafter for as long as the mare is receptive. On this schedule, motile sperm are present in the mare's reproductive system at all times when ovulation is likely to occur.

The ability to accurately predict ovulation has several important benefits. Significantly, only one live cover or insemination is needed to achieve a high rate of conception. Reducing the number of breeding encounters extends the stallion's semen usage, decreases the frequency of breeding accidents, and limits the transmission of sexual diseases. When many mares are being serviced, efficient breeding practices are almost essential if the farm is to be financially successful.

The two standard methods for determining when to breed are rectal palpation and ultrasonography.

RECTAL PALPATION

Ovarian changes that accompany heat, ovulation, and the development of the corpus hemorrhagicum are best appreciated through a series of examinations that involve palpation of the ovaries, cervix and uterus. To predict ovulation, palpation focuses primarily on the size of the ovarian follicles and their degree of firmness or softness.

Follicle development occurs in waves throughout the estrous cycle. Follicles are fluid-filled bubbles that grow within the ovary. As they enlarge, they project above the surface of the ovary, where they can be detected by feeling the ovary between the thumb and fingers. As heat approaches, palpation of the ovaries reveals many follicles 20 mm to 30 mm in size. Six days prior to ovulation, some follicles start to enlarge and become dominant. Dominant follicles grow at the rate of about 5 mm per day.

Eventually one follicle becomes larger than the others. This primary or ovulatory follicle, 35 to 50 mm in size, feels firm and tense. (Note that in 15 percent of mares, two follicles become primary, resulting in a double ovulation.) As ovulation approaches, the follicle becomes less tense and begins to soften; 24 hours before ovulation, the primary follicle becomes considerably softer, indicating that ovulation is imminent.

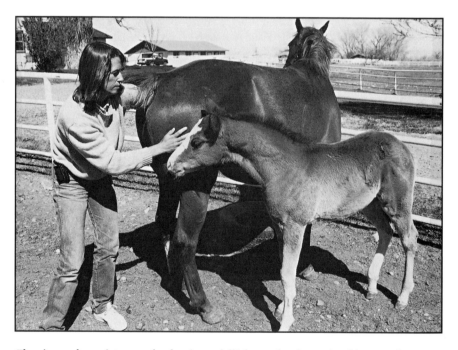

The size and consistency of a dominant follicle can be determined by rectal palpation.

For the average mare, palpation every other day beginning on the 3rd day of heat is sufficient to predict ovulation. Because palpation involves a measure of subjective interpretation, the experience of the examiner is an important part of the equation.

Ovulation occurs rapidly, with the follicle collapsing in a matter of seconds. The collapsed follicle can be felt as a crater or depression on the surface of the ovary at the site of the former follicle. After 8 hours, hemorrhage associated with the ovulation causes this cavity to fill with blood and serum. The result is a soft, mushy structure called the corpus hemorrhagicum (CH).

Over the next 2 to 3 days, the CH continues to fill with serum and blood. It may become tense and distended, in which case it can be easily confused with an ovulatory follicle. Any confusion between the two can be resolved by ultrasonography.

By 5 days postovulation, the CH has been replaced by the corpus luteum (CL). A mature CL is now located beneath the surface of the ovary and therefore is seldom detected by rectal palpation.

Palpation of the cervix and uterus also provides useful information about the estrous cycle. How firm the horns feel is referred to as uterine tone. In general, a horn with excellent tone is firm and tubular like a garden hose, while one with poor tone is flaccid like a collapsed segment of intestine. As the mare progresses from diestrus through early estrus to ovulation, uterine tone becomes less firm, progressing from good to fair or fair to poor. After ovulation, the process reverses.

The cervix is palpated through the rectal wall by compressing it against the floor of the pelvis with the tips of the fingers. In diestrus, the cervix is a long, firm, muscular tube about the length and width of an index finger. The cervical canal is tightly closed. Under the influence of estrogen, the cervix softens and relaxes, becoming half as long and twice as wide. The greatest degree of relaxation occurs just before ovulation. At this time, the cervical canal opens to permit the passage of sperm.

Cervical findings can also be noted on vaginal speculum exam. This examination, however, is not used very often because it adds no new information and exposes the mare's reproductive tract to air and contamination.

TRANSRECTAL ULTRASOUND

This noninvasive procedure uses high-frequency sound waves to produce a visual image on a computer screen. The portable ultrasound machine includes a probe or transducer, a computer and a monitor. The probe serves as both the source and receiver of the sound waves. Sound waves propagated by the transducer are carried into the body, where they encounter structures of differing density. High-density structures reflect back more sound than do low-density structures. The computer records the signals returning to the transducer and converts that information to a picture on the monitor in shades of white and gray.

Ultrasound frequencies are measured in megahertz (MHz). Low-frequency transducers (3 to 3.5 MHz) are best adapted to visualizing structures at greater distances from the tip of the probe. Higher frequency transducers (5 to 7.5 MHz) give better resolution to structures close to the probe.

The probe, held in the examiner's cupped hand, is inserted into the rectum. It then is moved from side to side to visualize the cervix, body of the uterus, horns of the uterus, oviducts and ovaries.

An ultrasound scan is slightly more accurate than rectal palpation for predicting follicle development and the time of ovulation. It is uniquely valuable for distinguishing the CH from the CL and for identifying double ovulations, twin pregnancies, anovulatory follicles and ovarian tumors.

In mares, the surface of the ovary is surrounded by a thick capsule of connective tissue that prevents follicles from continuing to grow outward. Accordingly, as follicles get larger, they are forced centrally toward the ovulation fossa. The ovulation fossa is the part of the ovary enclosed by the hooded end of the oviduct. Ovulation occurs within the ovary, and the egg passes directly into the ovulation fossa and tube.

Findings used to predict ovulation are the size and shape of the preovulatory follicle and the presence or absence of prominent folds in the endometrium of the uterus.

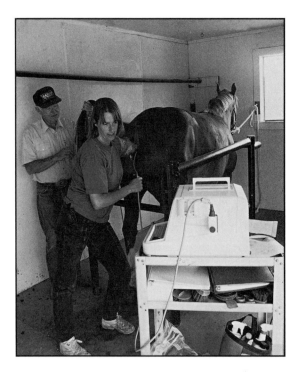

Transrectal ultrasonography is the most accurate method for predicting when ovulation will occur.

The size of the follicle is important because the likelihood of ovulation increases as the follicle surpasses 35 mm in size. Follicles grow at a rate of about 5 mm per day. Ovulation occurs, on average, at 45 mm (the range is 35 to 50 mm). During the 24 hours before ovulation, the largest preovulatory follicle migrates toward the ovulation fossa and can be seen to change configuration, going from a spherical shape to that of a teardrop or pear. Within 4 hours of ovulation, a tear in the wall of the follicle that points toward the ovulation fossa may be noted. As ovulation occurs, there is a sudden collapse of the large, fluid-filled follicle.

An ultrasound scan of the endometrium during estrus will show characteristic folds caused by estrogen-induced swelling. These folds become more prominent as estrus progresses, reaching their maximum 2 days before ovulation. Moreover, during the last 24 hours, the folds disappear rapidly, indicating that ovulation is imminent. These changes are best noted on serial scans.

HORMONE TESTS

Hormone levels for estrogen and progesterone correlate with the stage of estrous. Estrogen concentrations are highest just before ovulation and fall to baseline levels 2 days after ovulation. Progesterone concentrations are just the opposite. They are lowest just before ovulation and peak 6 days after ovulation.

Commercial tests are available for both hormones. However, these blood tests are expensive, and because of laboratory turnaround time, the results are usually not available in time to predict ovulation. Accordingly, hormone tests do not play an important role in determining when to breed.

PROBLEMS WITH DETECTING HEAT

Absence of heat is normal during winter anestrus when the ovaries are inactive. During the spring transition, a mare may develop a follicle and ovulate without showing signs of heat, although this is not common.

Failure to respond positively to the stallion during the natural breeding season raises the question of infertility and presents problems in determining when to breed. Fortunately, most mares have normal ovaries and cycle regularly—but some do not show heat for behavioral reasons (silent heat). When a non-behavioral cause is responsible, it is almost always a problem in the luteal phase of the estrous cycle (see PROLONGED DIESTRUS in chapter 2).

SILENT HEAT (BEHAVIORAL ANESTRUS)

Silent heat, also called psychological anestrus and unobserved heat, is a common problem characterized by normal estrous cycles but lack of receptivity to the stallion. The estrous cycle is considered to be normal if rectal palpation and/or ultrasound exams of the ovaries reveal growth and maturation of ovulatory follicles, and the serum progesterone is elevated during the 14 days of diestrus and low during the 7 days preceding ovulation.

Some silent-heat mares respond passively to the stallion, showing neither friendly nor unfriendly behavior. Most silent-heat mares, however, will exhibit diestrus behavior and actively reject the stallion.

Silent heat is a poorly understood phenomenon believed to have a psychological basis. Recent studies have shown that plasma estrogen concentrations in the period preceding ovulation are somewhat lower in mares with silent heat than in mares with normal estrous behavior. The significance of this is not known.

Note that mares vary extremely in how strongly they show heat to a stallion. Frequently, what appears to be lack of heat is simply failure to observe the more subtle signs of heat.

Silent heat is relatively common in maiden and nursing mares. Lactating mares are extremely protective of their foals and may not show heat until after the foal is weaned.

A common cause of silent heat is ineffectual teasing. When teasing is carried out on a regular schedule, maiden and nervous mares become comfortable with and attuned to the daily routine—often overcoming their shyness and ultimately showing heat. Random teasing, however, is counterproductive and often leads to missing the moment of receptivity.

SOME TECHNIQUES FOR ELICITING HEAT IN PROBLEM MARES

- *Obtain a teasing history.* Identify the problem mare and be prepared to spend an extra few minutes teasing each day. With intensive teasing, most mares in estrus will show signs of heat to the experienced handler.

- *Start teasing early in the breeding season.*

- *Change the teasing method.* To encourage intimate contact, remove the barrier and tease with both horses under control. Hobble the mare to prevent kicking. Allow the stallion to investigate the flank and vulvar areas.

- *Change the teasing location.* Switching sites can remove inhibitions and distractions associated with the initial site.

- *Change stallions.* The mare may prefer the second stallion to the first one, especially if the second stallion is less aggressive.

The administration of anabolic steroids containing derivatives of testosterone is another cause of silent heat. The effects are dose-related. At low doses, anabolic steroids suppress heat behavior but not estrous. At higher doses, however, they stop the estrous cycle. In addition to refusing to accept the stallion, steroid-treated mares show stallion-like behavior such as mounting mares in estrus and acting aggressively toward other horses. When steroids are withdrawn, it usually takes at least 6 months before the mare begins cycling and showing heat behavior.

Treatment of Silent Heat: Rectal palpation and/or ultrasound exams are used to determine when to breed. These exams are performed every other day until a preovulatory follicle is noted, and then every day until the mare ovulates.

PGF2α can be used to shorten diestrus and bring the mare into heat more quickly. This facilitates treatment by reducing the number of examinations that need to be performed.

A reluctant mare can usually be covered successfully when twitched and hobbled. The really difficult mare who flat refuses to breed despite twitching and hobbling presents a major difficulty for live cover. Artificial insemination is an excellent solution. Most breed registries in the United States allow the registration of foals conceived by fresh and cooled transported semen.

BREEDING ON THE FOAL HEAT

Breeding early in the calendar year is of great economic importance on many stud farms (see Operational Breeding Season in chapter 1). A policy common to the majority of breed registries in the Northern Hemisphere is to designate January 1 as the universal birth date. All foals born during the year automatically become yearlings on January 1 of the following year. The effect of the designation is to make foals born early in the year more valuable than foals born later in the year, since early foals are apt to be larger and stronger as 2-year-olds. As a result of this constraint, a mare who foals in late spring or summer may produce a less-valuable foal, and may not be a candidate for rebreeding until the following spring. The expenses of maintaining a barren mare, plus the potential for having fewer or less-valuable foals for sale, negatively affects the balance sheet.

The mare has a relatively long gestation, lasting on average 340 days. This means that a mare would have to become pregnant within 25 days after giving birth in order to produce a foal at about the same time next spring. The first postpartum heat occurs 5 to 12 days after foaling. The second postpartum heat occurs 31 days after foaling. Because the second heat would advance the birth date by about 1 month, there is considerable economic pressure to breed on the foal heat.

Early breeding is possible only because of the rapid rate at which the mare's uterus is able to shrink down (involute) and regenerate an endometrium conducive to the support of an embryo. On average, this occurs by the 15th day after foaling.

The mare is also unique in that her first postpartum heat is accompanied by ovulation. On average, ovulation occurs on day 10 after foaling, with a range of 7 to 20 days. After ovulation, there is a 5-day interval before the fertilized egg enters the uterus. So if conception occurred on day 10, the embryo would enter the uterus on postpartum day 15. In theory, breeding on postpartum day 10 or later should result in normal pregnancy rates.

In practice, fertility rates are 10 to 20 percent lower than normal when mares are bred on the foal heat. The reasons for this are that some mares ovulate before day 10, and others do not complete the process of uterine involution by day 15. In either situation, the uterus is not prepared to receive the embryo.

Ultrasound studies have shown that the likelihood of a successful pregnancy is directly related to the amount of residual fluid present in the uterus. Mares with little or no residual fluid in the uterus when the embryo arrives are less likely to suffer from early embryonic loss and more likely to have a successful outcome.

Based on these findings, many farms have adopted the policy of monitoring all postpartum mares with serial rectal exams and ultrasounds. If it does not appear that the mare will ovulate before day 10, and if she has little or no fluid present in the uterus, she is scheduled for breeding on the foal heat.

If the mare's uterus is not sufficiently involuted or she ovulates before day 10, she is not bred on the foal heat. However, time can still be saved by shortening the diestrus phase of the estrous cycle and advancing the second postpartum heat by about 1 week. This is accomplished by administering PFG2α 5 to 7 days after the mare ovulates.

An alternative to the above is to delay the first ovulation using hormonal therapy. Treatment is started empirically on the day of foaling. In one drug protocol, progesterone is given daily for up to 8 days, followed by a prostaglandin injection on the last day. In a second protocol, both estrogen and progesterone are given daily for 5 days. Both protocols delay ovulation so that conception occurs after postpartum day 15.

BREEDING

Mares can be bred by natural service or artificial insemination.

Natural service is also called live cover. It is the physical mounting of the mare by a stallion. This can take place in the setting of pasture breeding or hand breeding.

Artificial insemination involves the collection of semen in an artificial vagina and the infusion of that semen into the uterus of one or more mares. This subject is discussed in chapter 6.

PASTURE BREEDING

Pasture breeding involves putting a stallion out with a group of mares and letting nature take its course. The breeding is unrestrained and often unwitnessed. As a prerequisite for success, all individuals should have a breeding soundness examination before the breeding season to rule out sexually transmitted diseases.

The amount of acerage needed for pasture breeding depends on the number of mares. A large paddock is adequate for a stallion and a single mare. When a stallion is put out with a herd of 15 to 20 mares, it is best to have at least 40 acres.

Breeding management is limited to introducing new members to the herd and watching for problems. Since a stallion has a natural instinct to guard his herd, it can be difficult to handle individual mares on a regular basis to determine if they have become pregnant.

Pasture breeding provides opportunities for social interaction and courtship not found in hand-breeding programs. It is excellent training for young stallions, who learn the code of mating from experienced matrons. Mares with unexplained infertility may occasionally become pregnant when turned out with a seasoned stallion.

Pasture breeding is the best method if time, personnel, resources and facilities are limited. With adequate space, good forage, a fertile stallion and a small number of healthy resident mares, pasture breeding is cost-effective and produces pregnancy rates similar to those of many hand-breeding programs.

Pasture breeding provides opportunities for social interaction and courtship.

A major drawback of pasture breeding is the potential for injury to the stallion. Serious breeding accidents are infrequent, but there is always a possibility that a mare's kick to the penis or testicles could render a valuable stallion unserviceable.

HAND BREEDING

In hand breeding, a haltered mare is brought together with a haltered stallion, and the mating takes place with both horses restrained. Hand breeding requires facilities for boarding, a teasing program, and sufficient personnel to handle the horses.

Hand breeding has several advantages over pasture breeding. Because hygienic procedures are employed, there is less chance of transmitting venereal infections. As long as both horses are properly restrained, breeding injuries are not a factor. Mares who do not become pregnant on the first service can be rapidly identified and brought back to the stallion for rebreeding on the next estrus. The fertility of the stallion is easier to monitor; adjustments can be made in the stallion's book if the potency of his semen begins to decline with excessive use. Finally, hand breeding optimizes the stallion's use more effectively than pasture breeding. It is the only efficient way to provide natural service for a large number of visiting mares. Conception rates over 50 percent per cycle are common, with overall pregnancy rates of 85 percent reported on well-managed farms.

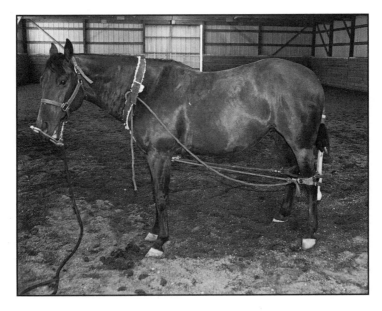

Hand breeding. For the safety of all concerned, the mare should be hobbled.

A lip twitch is sometimes used to calm the mare and encourage her to stand quietly.

PREPARATION

Before the mare is brought to the breeding shed, her readiness to breed should be reconfirmed by a positive teasing response to the stallion, as described in chapter 4. A change in teasing behavior from receptive to unfriendly indicates that the mare has ovulated. This can be confirmed by rectal palpation and/or transrectal ultrasound. If breeding cannot be accomplished within 12 hours after ovulation, it should not be attempted. It is extremely unusual for a mare to conceive when bred more than 12 hours after ovulation.

It is impossible to predict the behavior of a mare in the breeding shed. Even a mare with a sweet disposition is capable of acting violently if she panics or resents the stallion. Accordingly, for the safety of all concerned, it is essential to restrain all mares.

Breeding hobbles are commonly used to prevent kicking. While in hobbles, a mare can walk forward but cannot kick back at a stallion. Hobbles are often combined with a lip twitch to calm the mare and encourage her to stand still.

Breeding hobbles consist of a leather strap that buckles around the mare's neck in front of her shoulders. A rope extends from the lower part of this neck strap down between her front legs and attaches beneath her abdomen to another set of straps that buckle around the hocks. The connection between the front and rear restraints is secured with a quick-release knot.

The mare's hind shoes should be removed in preparation for breeding. As a further precaution, mares who are known kickers should be fitted with heavy felt kicking boots.

Some stallions bite down on the mare's neck during mating. This may result in a skin laceration and cause enough pain or fear to disrupt the mating. Mares should be protected by strapping on a leather neck and back shield, which the stallion can grasp and hold onto. Alternatively, the stallion can be fitted with a cage muzzle.

Small and maiden mares bred to a stallion with a large penis can incur tears of the cervix and vaginal canal. These injuries can be prevented by using a breeding roll to limit the depth of penetration. A breeding roll is a padded cylinder about 5 inches in diameter and 18 inches in length. It is covered by a sterile sleeve. As the stallion mounts, the roll is placed just above the stallion's penis between the stallion and the mare's buttocks.

Before the mare is brought to the breeding shed, her tail should be wrapped with a clean bandage, starting at the base of the tail and continuing to the end of the tailbone. The wrap minimizes contamination and prevents loose hair from being caught between the penis and vagina and lacerating the penis.

With the tail held to the side, the vulva and perineal area are thoroughly washed and rinsed with lukewarm water using paper towels, a soft fabric or a hand-held sprayer.

The purpose of the wash is to remove dirt and fecal material. It is not intended to sterilize the perineum. In fact, vigorous scrubbing and the use of disinfectants such as Betadine and chlorhexidine may actually be harmful. Studies have shown

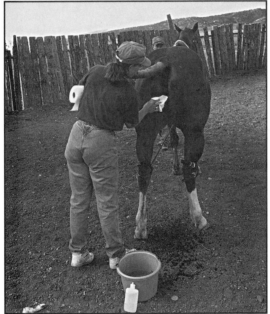

Washing the stallion's penis and the mare's vulva and perineum prior to breeding are important steps in preventing the transmission of venereal infections.

that washing with disinfectants may decrease the number of friendly bacteria and allow the colonization of pathogenic bacteria. If soap is used at all, it should be a mild one, such as Ivory Liquid. At the conclusion of the wash and rinse, paper towels are used to wipe the area dry, working from the vulvar lips outward in all directions. The mare is now taken to the breeding shed.

The stallion is brought to the breeding area and allowed to see the mare. Visual stimulation is usually sufficient for the stallion to drop his penis. The extended penis is gently washed and rinsed with lukewarm water. Betadine or chlorhexidine should not be used, because it irritates and inflames the skin of the penis.

A hand-held sprayer is ideal for washing both the mare's perineum and the stallion's penis. It eliminates the need for buckets and greatly reduces the potential for cross-contamination and the transmission of venereal diseases. If washbuckets are used, each horse should have its own bucket. The use of disposable plastic bucket liners eliminates the need to disinfect the buckets after each service.

COVERING THE MARE

The breeding shed should be large enough so that horses and handlers can move about freely. It is a good idea to pad the walls with a material such as urethane foam. The floor should provide good footing and be free of dust.

The procedures for hand breeding involve the services of two or three people. Each handler has specific duties and must know exactly what to do and what to expect at all times. The stallion and mare are controlled using a halter, chain and lead shank, as described in HANDLING THE STALLION, chapter 3. The chain is usually placed through the mouth or over the nose. This provides a degree of restraint that can be made more or less severe, depending on how forcefully the chain is pulled. Alternatively, a Chifney bit can be used.

The leads of both horses should be long enough so that if either horse rears, the handler can step back quickly without having to release the lead.

If the mare has a foal at her side, the foal should be confined in a holding pen or safely restrained by another handler within sight and sound of the mother.

The positions of the mare and her handler, and the stallion and his handler, are of critical importance for safety and control. *At no time should any handler stand between the two horses, or stand directly in front of or directly behind either horse.*

The mare handler stands at the mare's shoulder, usually on the same side as the stallion. The duties of the mare handler are to steady the mare, allow her to see the stallion, and prevent her from moving out from under the stallion as he mounts.

The stallion handler should stand at the left of the stallion's shoulder and lead him calmly and directly to the mare. An angled approach allows the mare to see and prepare for the stallion.

When an experienced stallion is presented to a calm mare in standing heat, he will usually greet her vocally, sniff and nuzzle her, display the *Flehmen response*, drop his penis and obtain an erection—all within about 3 minutes. A receptive

The horses are introduced.

Stand aside, but be prepared to control unruly behavior.

The position of the mare's tail prevents the stallion from entering.

The tail is held to the side.

Ejaculation occurs shortly after the stallion enters the mare.

mare, when suitably coaxed, spreads her legs, flexes her pelvis and strikes a breeding stance.

Although many horses do most or all of the above, the initial greeting does not always go this smoothly. Be prepared for the unexpected. As the two horses make contact, either may strike, rear, kick or bite. Mares in particular can strike with blinding speed. Both handlers must stand well out of the strike zone and be ready to instantly separate the horses.

Assuming that the initial greeting goes well, the stallion will remain at the mare's side until he is fully erect and ready to cover. This usually takes less than 15 seconds for the experienced stud but may take considerably longer for the novice stallion and the slow breeder. The stallion then raises up slightly to the side of the mare and quickly positions himself behind her buttocks for *intromission* and ejaculation.

A third handler (optional) stands beside the mare's left hip and holds the mare's tail to one side. Alternatively, the tail can be held by the stallion handler. It is sometimes necessary to guide the penis into the vagina, especially when the mare is relatively tall compared to the stallion. This step should be performed by the third handler. The third handler is also responsible for inserting a breeding roll between the stallion and mare above the stallion's penis. This handler should step back quickly after his or her duties are completed.

The period from intromission to ejaculation averages about 20 seconds. The normal ejaculatory pattern consists of five or six intravaginal thrusts, followed by three to five shorter thrusts accompanied by ejaculation. During ejaculation, the stallion's tail pumps up and down. This is known as flagging. Flagging, however, is not conclusive evidence that the stallion has ejaculated. Some stallions ejaculate but do not flag; others flag but don't ejaculate. You can often tell whether ejaculation is taking place by placing your hand under the stallion's penis and feeling for the pulsations. The most dependable way to verify that a stallion has ejaculated is to take a dismount semen sample as explained below.

As the stallion dismounts, the mare handler, who now stands on the mare's left side, quickly turns the mare's head to the left. This moves her hindquarters to the right and away from the stallion. At the same time, the stallion is backed off to the mare's left. These maneuvers help to prevent injury to the stallion in case the mare kicks on the dismount.

A semen sample can be obtained immediately after the stallion dismounts by milking fluid from the urethra into a clean cup. A drop of semen is placed on a glass slide and examined under a microscope. The presence of sperm verifies that the stallion has ejaculated. Although dismount samples are not reflective of the quality of the semen and are not used for fertility evaluation, finding blood, urine or white cells in the dismount ejaculate indicates the need for further investigation.

The stallion is now returned to the wash area, where his penis and sheath are thoroughly rinsed. Washing the stallion immediately after service is just as important in preventing venereal infections as washing him before.

ARTIFICIAL

INSEMINATION

GENERAL INFORMATION

Artificial insemination (AI) involves collecting semen from a stallion, mixing it with a semen extender, and inserting it into the uterus of a mare. Semen used within 12 hours of collection is called fresh. Semen that is cooled and maintained at 4° to 8°C (39° to 46°F) is capable of retaining viability for several days. Semen that has been frozen can be preserved for months or years if kept at a temperature of –196°C.

Most breed registries in the United States allow the registration of foals conceived with fresh or cooled semen. However, most of them specifically prohibit the use of frozen semen. These policies are subject to change. Be sure that you know the current regulations if you intend to register a foal conceived by AI.

Breeding by AI, especially using transported cooled semen, has several advantages over breeding by natural service:

- AI efficiently uses the stallion's reproductive capacity. A single ejaculation can be divided into several insemination doses and used to breed several mares.

- AI gives owners of broodmares access to superior stallions across the country.

- It eliminates the risk of a breeding accident harmful to a valuable stallion or mare.

- It greatly reduces the risk of acquiring a venereal disease.

- It provides an excellent alternative when live cover is impossible because of a physical impediment or the unwillingness of the mare to stand for the stallion.
- It eliminates the expenses associated with transporting and housing the mare (and often her foal) at the breeding farm.

The principal drawback of AI is that it requires personnel experienced in the collection, handling and storage of semen. When using cooled-transported semen, the mare must be monitored closely to ensure that the semen is ordered and delivered at the optimum time for insemination. This requires close cooperation between those responsible for collecting the semen and those responsible for inseminating the mare.

Finally, the semen of some stallions does not survive the cooling process well. This results in a high rate of failure and diminishes the value of these stallions for breeding with cooled-transported semen.

PREGNANCY RATES

Per-cycle pregnancy rates using artificially inseminated fresh semen approach 80 to 90 percent. These rates are equivalent to (and in some studies even superior to) those obtained by natural service.

Rates obtained with cooled semen vary with the individual stallion and how well his semen tolerates cooling and thawing. Even more important is how long the semen is stored. For example, mares inseminated with good-quality semen stored for 24 hours typically achieve per-cycle pregnancy rates in the range of 60 to 70 percent. If the semen is stored for 48 hours, however, the pregnancy rates drop to 40 to 50 percent.

Pregnancy rates for frozen semen are discussed below.

SEMEN COLLECTION

Most stallions can be trained to use an *artificial vagina* (AV). Three standard models are available. All consist of an outer cylinder about 2 feet long surrounding a hollow inflatable bladder. The bladder is filled with warm water immediately prior to use. Inside the bladder is a combination liner and cone, usually of disposable plastic. The cone connects to a container for collecting the semen. An optional in-line nylon micromesh filter in the AV or collection bottle is available.

Many stallions can be trained to mount a phantom mare for semen collection. Some stallions require the presence of an estrus mare adjacent to the phantom.

A phantom mare can be purchased commercially or made by welding two hot-water heaters together and mounting them on steel pipes bolted to a cement pad. The frame then is padded with several layers of foam and covered with vinyl.

Artificial vaginas (left to right): CSU Model, Missouri Model, and Japanese Model. Photo: *Equine Reproduction,* Angus O. McKinnon, BVSc, and James L. Voss, DVM, eds., Lea & Febiger (1993). Reproduced by permission.

A stallion who has not been trained to mount a phantom can be collected from a live or *jump* mare. A jump mare is a mare in heat, or a mare whose ovaries have been surgically removed. A neutered mare is often given estrogen to enhance the signs of heat.

Most breeders prefer the convenience of a phantom for semen collection. The procedure is easier, the semen quality is more consistent, and the potential for injuries during collection is greatly reduced.

The stallion is brought into the breeding shed and shown the phantom or live mare to obtain an erection. The penis is washed and rinsed with warm water. The AV bladder is inflated with warm water at the temperature preferred by the stallion (about 110° to 115°F). This is best determined by a thermometer. A sterile, nonspermicidal lubricant such as H-R™ jelly is used to lubricate the AV.

If an estrus or jump mare is used, she should be prepared exactly as described for covering the mare. This involves wrapping her tail, washing and rinsing the vulva and perineum, and applying breeding hobbles and sometimes a lip twitch. The same handling precautions should be used as described in COVERING THE MARE in chapter 5. The potential for injury to horses and handlers is the same in both situations. The person holding the AV must be prepared to move away quickly as the stallion begins to dismount.

A phantom mare constructed from hot-water heaters.

The frame is padded and covered with vinyl.

The urine from an estrus mare can be poured over the phantom
mare to enhance its appeal to the stallion.

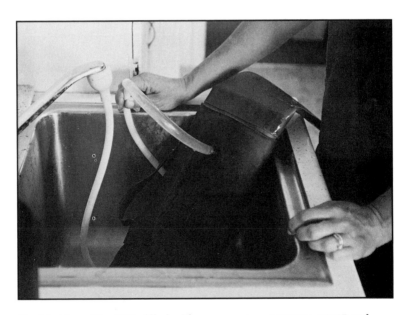

The bladder of the AV is filled with warm water at 110° to 115°F and
lubricated with K-Y™ jelly.

The stallion is led up to the phantom and displays the Flehmen reaction in response to the smell of the mare's urine.

The stallion mounts.

The AV is positioned at the correct angle for the stallion's penis and tilted so that the semen runs into the container.

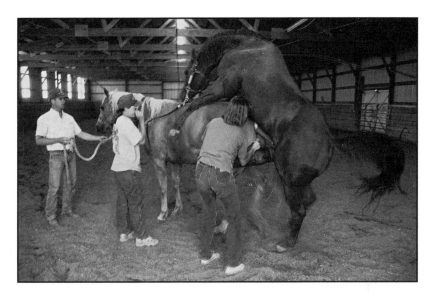

Collecting semen from a jump mare requires the same precautions as described for hand breeding.

The stallion is now allowed to mount the phantom or jump mare. The handler responsible for collecting the semen is positioned on the left side of the stallion. The AV is held at an angle that approximates the angle of the stallion's erection. The weight of the AV is supported by grasping the handle with the left hand and holding it against the phantom or the mare's left side with the hips. Ejaculation should occur in less than 30 seconds. At the conclusion of the procedure, the stallion's penis should be washed with warm water to remove all remaining lubricant.

SEMEN HANDLING

The semen should be transported to the laboratory within minutes of collection. If an analysis is planned, all equipment and supplies, including the semen extender, should be prewarmed to body temperature (see SEMEN ANALYSIS in chapter 3).

If an in-line filter was not used during collection, the semen should be passed through a nontoxic filter to remove the gel, or the gel fraction can be aspirated with a syringe.

The semen is diluted with semen extender to improve sperm cell survival.

Regardless of whether the semen will be used immediately or later on, it should be mixed with a high-quality prewarmed semen extender to improve the chances of survival between collection and insemination. Semen extenders are composed of lipoproteins such as those found in milk. Glucose is often added to provide energy for cell metabolism. One or more antibiotics are added to prevent or retard bacterial growth. Semen extenders can be homemade or purchased commercially.

The semen should be diluted with the extender at a ratio of at least 1 part semen to 1 part extender. In some cases, the ratio may be higher. Studies show that sperm survival is maximized when the semen is diluted to a final concentration of 25 to 50 million sperm per mL.

Subsequent handling depends on whether the semen will be used fresh, cooled or frozen. Fresh semen can be stored for up to 12 hours at room temperature, as long as it is protected from light. It should not be stored in an incubator at body temperature (37°C), because this will destroy the sperm.

INSEMINATING THE MARE WITH FRESH SEMEN

INSEMINATION VOLUME

The total number of motile sperm in the insemination dose determines the fertility of the dose. Experience indicates that doses containing 300 to 500 million progressively motile sperm maximize the chances for pregnancy. Greater numbers do not increase pregnancy rates, but numbers below 100 million significantly decrease pregnancy rates. Allowing for handling errors, a standard insemination dose of 500 million progressively motile sperm is considered optimum when inseminating fresh semen. For cool-stored semen in which motility after 24 hours will be reduced by one-half or more, the standard insemination dose is no less than 1 billion.

The volume of extended semen needed to deliver the required number of sperm depends on the concentration of sperm. The volume is determined by dividing the concentration of progressively motile sperm per mL of gel-free ejaculate by the total number of sperm desired. Thus if the concentration of progressively motile sperm is 50 million sperm per mL, the volume of semen required to deliver 500 million sperm would be 10 mL.

Typical insemination volumes for extended semen range from 30 to 50 mL. When using smaller volumes, precautions must be taken to ensure that a significant fraction of the insemination dose does not remain in the container and pipette.

TIMING AND FREQUENCY OF INSEMINATION

Most stallion managers cover the mare beginning on the 3rd day of standing heat and then every other day thereafter until she goes out of heat. On this schedule, motile sperm should be present in the mare's oviducts at all times when ovulation is most likely to occur.

INSEMINATING

The procedure is performed under sterile precautions using plastic disposable equipment. The mare's tail is wrapped and held to the side. The mare's vulva, clitoral fossa and perineum are repeatedly washed using a mild surgical soap such as 2 percent chlorhexidine (Nolvasan). It is important to thoroughly remove all soap by spray or rinse, because residual soap is spermicidal.

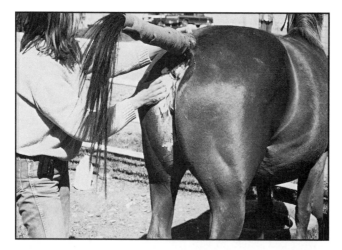

Inseminating the mare. The tail is wrapped and the perineum washed to prevent contamination.

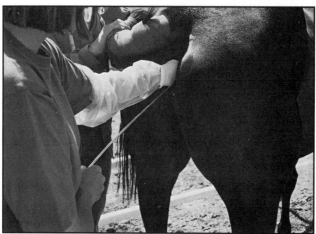

The insemination pipette is guided into the uterus.

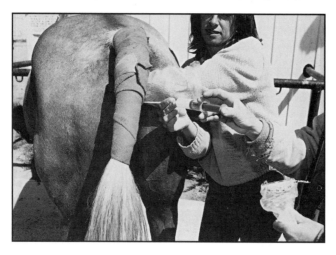

The measured amount of semen is slowly injected and deposited into the uterus.

An 18- to 20-inch sterile plastic disposable insemination pipette is guided with a sterile gloved hand through the cervical canal and into the body of the uterus. The insemination dose is drawn into the plastic syringe. The syringe is attached to the pipette, and the semen is slowly deposited into the uterus.

Note that the tip on the plunger of some syringes is made of rubber rather than plastic. Rubber may be toxic to the semen of some stallions after as little as 1 minute of contact. Therefore it is preferable to use syringes with plastic-tipped plungers.

INSEMINATING WITH COOLED-TRANSPORTED SEMEN

The critical factor in breeding with cooled semen is loss of sperm motility (and therefore fertility) with the passage of time. A typical ejaculate might have an initial motility of 65 percent, but only 45 to 50 percent after 24 hours. At 48 hours, the motility drops to 30 percent. Because sperm lose motility rapidly, the number of inseminations from a single ejaculate is usually limited to two, assuming that the semen is received within 48 hours of collection (see ORDERING THE SEMEN on the following page).

TESTING THE STALLION

Not all stallion's semen survives the cooling process. Frequently the semen of a stallion quite potent on the breeding farm simply does not "ship well." When selecting a stallion as a semen donor, it is important to have some idea of his success rate with cooled semen. What percentage of mares become pregnant? How many deliver live foals?

Unfortunately this information usually is not available. A stallion who has seen limited service with cooled-transported semen won't have meaningful statistics. Furthermore, the handling of semen as well as the timing of insemination are variables that greatly affect outcome.

There are, however, some questions you can ask that will give you an indication of the professionalism of the breeding farm. For example, will a semen analysis be supplied with the shipment? Does the stallion owner guarantee the viability of the cooled semen on arrival as judged by your veterinarian? What are your recourses if the mare does not become pregnant? If you are not satisfied with the explanations and guarantees given, you should consider using a different stallion.

The only way to determine if an untested stallion's semen will survive the storage process is to test the semen under conditions similar to those of actual shipping.

The semen is collected and processed as described earlier in this chapter. It then is placed in a shipping container designed to maintain a temperature of 4° to 8°C (39° to 46°F). Because the characteristics of the container dictate the cooling rate and storage temperature, it is important to place the semen in the same container that will be used to ship the semen.

After 24 hours, a sample of the cooled semen is removed from the container and warmed to 37°C. Sperm motility is evaluated after 15 minutes of warming. If sperm motility is acceptable after 24 hours of cool storage, it is relatively certain that the stallion's semen will survive the cooling process, and the stallion can be recommended for breeding with cooled-transported semen.

ORDERING THE SEMEN

With shipped semen, pinpointing ovulation is essential. Best results are obtained when the mare is inseminated 12 to 24 hours prior to ovulation (see PREDICTING OVULATION in chapter 4). The owner of the mare must coordinate with the stallion manager to ensure that semen is ordered 1 to 2 days before the mare is expected to ovulate. Ideally the semen will arrive just a few hours before ovulation and can be used at once.

It is preferable to have the semen arrive too early rather than too late (after the mare has ovulated). If the semen arrives and it appears by rectal palpation or ultrasonography that the mare is not yet ready to ovulate, ovulation can be induced with HCG, as described in chapter 2. The first insemination is given along with the HCG and the second 12 hours later. For this plan to be successful, the mare should have a mature follicle 35 mm to 40 mm in size.

SHIPPING THE COOLED SEMEN

Shortly after collection, the semen should be placed in a container that cools it at a relatively slow rate to 4° to 8°C (39° to 46°F). The most popular container for this purpose is the Equitainer™ system by Hamilton-Thorn (Danvers, Massachusetts). The container holds two coolant packets that have been frozen. The design of the container allows extended semen to cool slowly and remain at the desired storage temperature for up to 60 hours. Disposable shipping containers also are available.

The semen is usually packaged in plastic containers. Baby-bottle liners are ideal for this purpose. Two insemination packets are normally placed in each container. This allows the mare owner to use one dose immediately and the second 12 to 24 hours later.

The container is sealed and dispatched by a carrier that guarantees delivery in 24 hours, such as UPS or Federal Express. In some parts of the United States (particularly in rural areas), Saturday deliveries may not be possible. Furthermore, most commercial carrier services will not pick up or deliver on Saturdays, Sundays and holidays. If arrangements are made to collect and ship the semen on a weekend or holiday, the container will need to be hand-carried to the local airport and shipped counter-to-counter.

Two insemination packets in plastic baby-bottle liners ready for shipment.

The Equitainer system cools and stores the semen at the desired temperature for up to 60 hours.

USEFUL INFORMATION TO INCLUDE
WITH THE SEMEN SHIPMENT

- Name, address and telephone number of the stallion owner and the veterinarian responsible for collecting and processing the semen.
- Stallion identification—name, breed, registration number, markings and brands.
- Semen information—date and time of collection, semen extender and ratio used, volume of extended semen shipped, number of sperm per mL of extended semen, total number of sperm shipped, and initial sperm motility.
- Paperwork required by breed registry.

Most stallion owners send an information sheet along with the shipment. Ideally, this form includes the ratio of semen to extender, as well as test results on the semen.

INSEMINATING

Do not open the semen container until your veterinarian is present and all preparations are completed for the insemination.

The procedure is the same as that described for fresh semen. It is not necessary or advisable to warm the semen prior to insemination.

For quality-control purposes, it is a good practice to hold back a small fraction of semen and have your veterinarian analyze it for percent motility. The fraction is warmed to 37°C for 15 minutes, and the motility is estimated as described in SEMEN ANALYSIS, chapter 3.

If the motility is poor, the stallion owner should be notified, because this may represent an unexpected finding. If on repeat breeding, the motility is consistently poor and the mare fails to conceive, an adjustment may be made in the stud fee or another stallion offered as a substitute.

When a stallion's fresh semen has acceptable motility but his cooled semen does not, the semen can be tried using different semen extenders, dilution ratios or shipping containers. This may result in finding a combination that significantly improves the quality of his cooled transported semen.

FROZEN SEMEN

The principal advantage of frozen (cryopreserved) semen is that, using a liquid nitrogen container, it can be shipped anywhere in the world, thus greatly expanding the population of mares who can be bred to a specific stallion. Another major advantage is that frozen semen can be stored indefinitely and used to preserve the breeding potential of a valuable stallion long after his death.

Pregnancy rates achieved by breeding with frozen semen are lower than those attained with fresh or cool-transported semen. There is extreme variation in the ability of semen to survive freezing, not only among individual stallions but also among ejaculates from the same stallion. The reasons for this are unknown.

When stallions are selected at random, per-cycle pregnancy rates using cryopreserved semen will range from 8 to 61 percent. The selection of stallions whose semen freezes favorably, as well as the use of different protocols for freezing and thawing, undoubtedly contributes to higher rates in some series.

However, if you consider only those stallions whose initial post-thaw sperm motility is 40 percent or better (about 30 percent of stallions), per-cycle pregnancy rates of 30 to 50 percent are possible. Rates based on two or more insemination cycles are somewhat higher.

Freezing of semen requires special procedures that involve the use of various cryopreservatives along with specific rates of freezing. All factors—including the extender, cryopreservative, method of packaging, and rates of freezing and thawing—must be considered in total since they all interact to protect the sperm during critical stages.

Packaging systems for freezing sperm include pellets, plastic and vinyl straws, and aluminum packets. Different packaging systems alter the percent motility and pregnancy rates for the semen of some stallions, but there is no universal packaging system currently recognized as superior to all others. The individual packaging system does, however, influence the rate at which the semen should be thawed. Time and temperature must be carefully controlled. Even a small margin of error can lead to the death of sperm.

For the stallion whose sperm does not freeze well, trying his sperm under a variety of conditions using different freezing protocols might make it possible to identify an extender, cryopreservative or packaging system that will significantly improve the ability of his sperm to tolerate freezing and thawing.

Because of the many technical aspects of cryopreservation, it is advisable to take the stallion to a university or equine reproduction center that specializes in the collection, processing and storing of frozen semen.

Timing of insemination with frozen-thawed semen is similar to that for cool-stored semen. Thawing techniques and the total number of sperm vary with the packaging system and the protocol used. The veterinarian responsible for inseminating the mare should be familiar with the specific protocol for handling, thawing and reconstituting the semen received.

PREGNANCY

GESTATION

Gestation is the period from conception to birth. As calculated from the first day of breeding, it averages 340 days, with a range of 327 to 357 days.

An extra long gestation, however, is not uncommon in mares. Normal pregnancies can last as long as 399 days. Apparently the mare has some ability to adjust the length of pregnancy so that her foal will be born with the best chances for survival. Delayed embryonic development and hereditary predisposition are other possible explanations for this syndrome of prolonged gestation. Foals born of long gestations are not oversized and do not cause difficult labor. Despite concern about the length of gestation, if the foal is moving about normally in the uterus and there is no udder enlargement or signs of milk production, it is better to wait until the mare is ready to deliver rather than attempt to induce labor.

A foal born alive before 320 days is considered premature. A foal born before 300 days is usually too immature to survive.

DETERMINING PREGNANCY

TEASING RESPONSE

One of the earliest signs of pregnancy is failure of the mare to tease back in or exhibit heat 17 to 21 days after service. Teasing is an excellent way to screen for pregnancy; however, the results of teasing should not be taken as conclusive evidence of pregnancy or the lack thereof. As noted in chapter 4, "Determining When to Breed," teasing is only 80 percent accurate. Silent-heat mares, for example, don't display heat and will appear to be pregnant even though they are not.

Furthermore, an occasional mare who is pregnant will show heat to the stallion and thus appear not to be pregnant.

Furthermore, a number of conditions (other than pregnancy) are accompanied by an elevated serum progesterone and a negative teasing response. They include early embryonic death, ovulation in diestrus, persistent corpus luteum and pseudopregnancy.

PALPATION PER RECTUM

This is a simple and inexpensive method of determining pregnancy. The principal drawback of rectal palpation is that the ability to detect pregnancy *before 30 days' gestation* is highly dependent on the skill of the practitioner.

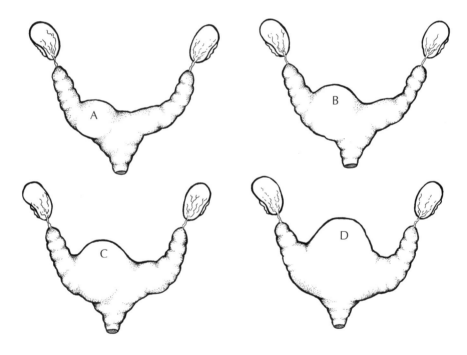

Findings on palpation of the uterus during pregnancy: (A) Early pregnancy vesicle at 20 days' gestation. (B) At 30 days' gestation. (C) At 40 days' gestation. (D) At 60 days' gestation.

By 20 days' gestation, the *tone* of the uterine horns is excellent and the cervix is closed, narrow and elongated like a pencil. A slight bulge, the embryonic vesicle, can sometimes be felt at the base of one of the uterine horns. The *conceptus* is about 3 to 4 cm.

Palpation at 30 days' is more accurate. The embryonic bulge is 4 to 5 cm. The wall of the uterus is somewhat thinner at the vesicle, and there may be a distinct impression of fluid within the bulge. The tone of the uterus remains excellent. The cervix is firm and contracted.

By day 40, the fetal-fluid bulge is 6 to 7 cm and has a fluid volume of about 200 mL. The tone of the uterus now becomes slightly less firm.

At 60 days' gestation, the bulge is 10 to 13 cm.

At 90 days', the uterus has been pulled over the brim of the pelvis into the abdomen. This makes it difficult to estimate the age of the fetus by palpating the size of the uterine horn. However, by 6 months' gestation, the fetus occupies the body of the uterus and can now be palpated without difficulty.

SERUM PREGNANCY TESTS

Equine chorionic gonadotropin (eCG): There are three serum pregnancy tests based on detecting eCG, a hormone manufactured by the placenta between days 37 and 100 of gestation. These tests are the *Mare Immunological Pregnancy* (MIP) test, *Direct Latex Agglutination* (DLA) test, and the Enzyme-linked Immunosorbent Assay (ELISA) test. The MIP is the one used most often, but all three assays are accurate after 50 days' gestation.

eCG is manufactured by the endometrial cups, which develop at 37 days' gestation. Levels begin to rise after day 40, peak at about day 60, and begin to decline thereafter. After the endometrial cups develop, they will continue to produce eCG for up to 2 to 3 months—even if the pregnancy is lost. Accordingly, eCG tests will not detect a pregnancy before 40 days' and can give a false positive reading between 40 and 120 days of gestation.

Progesterone: An elevated serum progesterone level on the 16th day following ovulation is suggestive of pregnancy and is 90 percent accurate if the level remains elevated. A few nonpregnant mares will maintain high progesterone levels as a result of a persistent corpus luteum, as discussed in chapter 2.

ULTRASONOGRAPHY

Transrectal ultrasonography is 95 percent accurate for pregnancy at 15 days' postovulation. In fact, using a high-frequency transducer (7.5 or 5 MHz) and a high-quality monitor, a small, round pregnancy vesicle can often be seen in the uterine body as early as 10 days' postovulation.

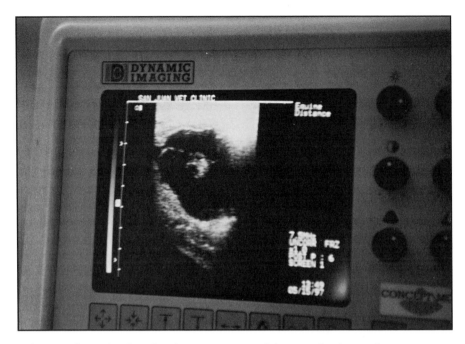

Ultrasound examination showing pregnancy vesicle at 14 days' gestation.

To avoid unnecessary and costly screening, most veterinarians prefer to scan mares at 14 to 18 days'. Screening at this time excludes most nonpregnant mares who will have already returned to estrus, leaving those who are likely to be pregnant for screening.

However, two very important groups of mares who should be screened before 16 days' gestation are Thoroughbreds and other breeds with a known tendency to ovulate more than one follicle, and all mares with a history of twinning or multiple ovulations. Recall that equine embryos are highly mobile within the uterus until the 16th day. The diagnosis of twin pregnancy before the vesicles become fixed makes it easier to selectively terminate one pregnancy, as described in Twin Pregnancy (chapter 9). This greatly simplifies the management of twins.

Although the embryonic vesicle can be detected by 15 days' gestation, the embryo within the vesicle cannot be detected before day 20. The embryonic heartbeat, an important indicator of fetal viability, can be identified by day 22. An absent heartbeat may indicate that the embryo is dead, even though the vesicle continues to grow. The ultrasound scan should be repeated in 2 to 3 days. If embryonic death is confirmed, pregnancy can be terminated with prostaglandin PGF2α, enabling the mare to return to estrus in 3 to 5 days.

The best time to determine the sex of the foal by ultrasonography is between 60 and 80 days' gestation.

CARE OF THE PREGNANT MARE

FEEDING DURING PREGNANCY

During the first 8 months of pregnancy, continue to feed the mare her usual ration. There is no need to change the composition of her ration or to increase the amount of feed until the last trimester. Trace-mineralized salt should be available free-choice. In selenium-deficient areas, use trace-mineralized salt containing selenium, and consider giving a vitamin E supplement as well.

Water requirements increase greatly after midpregnancy. The mare should have free access to clean, fresh water at all times.

During the last 3 months of gestation, there is a significant increase in the size and weight of the foal. This creates additional energy and protein needs. These needs should be met by increasing the quantity of the mare's feed so that she is gaining about 1 pound a day and appears well-conditioned to moderately fleshy, with a body-condition score (see appendix) of 5 to 6.

The nutrient requirements of an 1,100-pound mare in late pregnancy are given in Table I, appendix. In essence, the mare will require 15 percent more dietary energy and protein and about twice as much calcium, phosphorus and vitamin A.

A mare in late pregnancy requires 9 to 10 percent protein in her total daily ration. If she is consuming average or better quality alfalfa-legume or grass hay (either as the sole source of feed or in a hay-to-grain ratio of 70 to 30), no additional protein should be required. These diets should provide enough energy and protein to meet all of her needs during late pregnancy and throughout lactation— provided that ample amounts of feed are available for free-choice consumption.

However, when the mare is consuming pasture or grass hay cut at a late stage of maturity and containing only 8 percent protein, and in addition is consuming grain in a hay-to-grain ratio of 70 to 30, the grain mix should contain at least 14 percent protein. A customized grain mix can be formulated using soybean meal as the protein supplement, or a commercial grain mix can be used. Commercial mixes containing 16 percent protein, 0.8 percent calcium and 0.6 percent phosphorus are highly suitable.

The mare should be fed at a rate to maintain a body-condition score of 5 to 6. The amount to feed depends on the quality of the feedstuffs and the condition of the mare. As an example of the average range, a 1,200-pound late-pregnancy mare can be expected to consume 13 to 16 pounds of hay and 4 to 5 pounds of grain per day.

Calcium and phosphorus requirements increase in late-pregnancy. Hay and forages usually cannot provide adequate amounts to meet these demands. This might not be a problem if the mare has ample body reserves. However, with several consecutive pregnancies, these reserves can be depleted. Accordingly, it is a good policy to provide a supplemental source of calcium and phosphorus. Your veterinarian can recommend an appropriate mineral prescription based on knowledge of local conditions or analysis of the calcium and phosphorus content of your hay and grain. Ration analysis is a service provided by most feed stores and mills.

Vitamin A (and its precursor, carotene) is the only vitamin that might not be present in adequate amounts in rations routinely fed to horses. The best sources of carotene are rapidly growing spring and early summer grasses. Horses who consume such forage for 4 to 6 weeks store enough vitamin A to maintain adequate serum levels for 3 to 6 months. Properly harvested, early-cut legume hay (such as alfalfa) has adequate amounts of carotene. However, hay stored for 6 months loses half its carotene content.

To prevent vitamin A deficiency when poor-quality hay or forage has been consumed for long periods, add 30,000 IU vitamin A to the daily diet during the last 90 days of pregnancy and throughout lactation. Vitamin A supplements are available at feed stores. Alternatively, a vitamin A injection given by your veterinarian will provide adequate vitamin A reserves for at least 3 months.

Vitamin E (along with selenium) enhances the body's immune system. Vitamin E is present in green forages and natural feeds in amounts that more than meet the daily nutritional requirements of pregnant mares. A deficiency is conceivable if a mare were fed pelleted feeds and/or cured hay for an extended period. Vitamin E deficiency in the mare has been implicated as a cause of equine degenerative myeloencephalopathy in foals. Wheat-germ oil and alfalfa meal are excellent sources of supplemental vitamin E.

Selenium-deficient diets produce foals with nutritional myopathy, a disease characterized by weakness, difficulty with suckling and swallowing, respiratory distress and heart failure (see chapter 15, "Care of the Newborn Foal"). In selenium-deficient areas, use trace-mineralized salt containing higher levels of selenium.

The microminerals iodine, zinc and copper may be deficient in some forages, hays and grains. These deficiencies will not occur if the mare is allowed free access to trace-mineralized salt.

There is no evidence that oversupplementing the pregnant mare with vitamins and minerals is beneficial. In fact, it may well be detrimental. For example, giving excess calcium, phosphorus, or vitamin D causes abnormal bone development and flexor tendon contracture in foals. Excess iodine in the form of extracts of seaweed causes retarded muscular development and neonatal goiter. Selenium excess produces toxicity characterized by loss of hair from the mane and tail and sloughing of the hooves. Accordingly, vitamin and mineral supplements should not be used unless to correct a proven deficiency and then only under veterinary guidance.

EXERCISE

A physically fit mare is better adapted to delivering a healthy foal and is less likely to have postpartum complications. Moderate exercise maintains muscle tone and condition. The mare can be ridden for pleasure up to the last month of gestation. However, intense physical exertion should be avoided. Overexertion has been linked to pregnancy loss.

Mares on pasture get adequate exercise. Those confined to a stall or small pen should be exercised twice daily.

VACCINATIONS

The vaccination schedule for pregnant mares is shown in Table II, appendix. Keeping the vaccinations current ensures that high levels of antibodies will be present in the mare's colostrum.

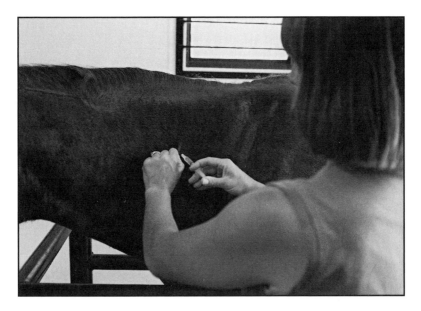

Vaccinations should be given 3 to 6 weeks before foaling to achieve maximum protection for the foal.

Following exposure to the rhinopneumonitis virus, there is a significant risk of abortion. Because immunity following rhinopneumonitis vaccination is of short duration, vaccinations should be given during the 5th, 7th and 9th months of gestation in order to ensure high levels of protection. Some practitioners recommend vaccinating mares during the 3rd month as well. A modified live and a killed vaccine are available. Pregnant mares should be vaccinated only with the killed vaccine.

The other vaccinations recommended during late pregnancy are tetanus toxoid, Eastern and Western equine encephalomyelitis, and equine influenza. To achieve maximum protection for the foal, these vaccinations should be given 3 to 6 weeks before foaling.

INTESTINAL PARASITES

All horses, including pregnant mares, should be on a routine deworming schedule using agents that are effective against the large and small strongyles, bots, ascarids, pinworms and threadworms. Intestinal parasites drain the mare of nutrients, migrate and damage organs, lower resistance to infection, and may harm the developing fetus.

Several deworming programs and anthelmintics are available. Which agents to use and when to use them will depend on local conditions as determined by your veterinarian. Because of the risk of causing abortion, organophosphate dewormers, such as those containing Dichlorvas or trichlorfon, should not be used during pregnancy. In addition, benzimidazole dewormers are not recommended during the first 4 months of pregnancy.

The anthelmintic drug ivermectin is effective in treating nearly all species of intestinal parasites and is safe to use in pregnant mares and newborn foals. It is given as an oral paste once every 2 months. Despite its frequent use, drug-resistant worms have not emerged.

On farms where threadworms have been a problem, mares should be given ivermectin or oxibendazole within 24 hours of foaling to prevent the passage of threadworms in the milk.

PREPARING FOR FOALING

If the mare is not going to foal at home, she should be taken to the place where she is going to foal at least 6 weeks prior to foaling. This allows her to adjust to the surroundings and become familiar with the new routine. Transporting a mare over a long distance can cause severe emotional and physical stress. When possible, long trips should be avoided during the last 2 months of pregnancy.

The expectant mare can be kept in a large paddock or exercise pen during the day, but should be put in her stall box at night to become accustomed to the area where she will foal. The best bedding for a foaling stall is clean straw, although wood shavings are satisfactory.

A mare with a Caslick's operation for *pneumovagina* (see chapter 8) should have the vulva surgically opened to prevent tearing of the perineum during foaling. This should be done 30 days before her expected date of delivery.

Remove the mare's horseshoes before she foals.

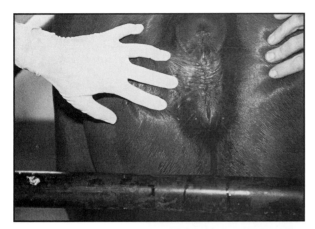

A Caslick's vulvoplasty should be opened 1 month before foaling.

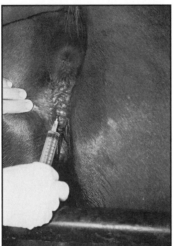

A local anesthetic is injected to numb the area.

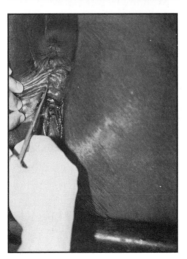

The scar tissue is divided with scissors. Bleeding is minimal.

PREGNANCY COMPLICATIONS

HYDROPS OF THE FETAL MEMBRANES

The fetus is surrounded by two fluid-filled sacs formed by the fetal membranes. The outer sac is called the allantochorion. Inside the allantochorionic sac is a second, smaller sac called the amnion. The amnion surrounds the fetus.

Hydrops of the fetal membranes is a rare, potentially fatal condition that occurs after 7 months' gestation, in which a huge amount of fluid accumulates in the allantochorionic sac, the amniotic sac, or both sacs. The cause is unknown, but many cases are associated with fetal malformations.

Signs are due to rapid enlargement of the uterus. Over 10 to 14 days, the mare exhibits abdominal swelling, colic, respiratory distress, difficulty walking, and may eventually be unable to stand. Complications of hydrops include rupture of the prepubic tendon and/or abdominal wall muscles, rupture of the uterus and spontaneous abortion.

Treatment: Pregnancy must be terminated to save the life of the mare. This is accomplished by gradually dilating the cervix and rupturing the membranes. The fetus must often be manually extracted. Intravenous fluids are needed to replace the large volume of fluid lost following the release of the hydrops.

RUPTURED PREPUBIC TENDON AND ABDOMINAL WALL MUSCLES

The prepubic tendon attaches to the pubic bone and serves as a common tendon for the muscles of the abdominal wall. The weight of the pregnant uterus may cause this tendon to give way gradually or separate suddenly. Sudden, complete rupture is accompanied by shock, collapse and death.

A similar rupture of the abdominal wall muscles without inolvement of the prepubic tendon can occur as a separate event.

Older draft mares and fat idle mares are at highest risk for both conditions. Signs include abdominal pain, reluctance to walk, the accumulation of fluid beneath the skin of the midline of the abdomen and a pronounced sagging of the belly.

Treatment: The immediate induction of labor with delivery of the foal is the best treatment. However, this might not be possible, owing to prematurity (see DETERMINING FETAL MATURITY in chapter 13, "Foaling"). In that case, restrict exercise, confine the mare and support her abdomen with a sling. As soon as the foal is mature, your veterinarian can induce vaginal delivery or deliver the foal by cesarean section. Techniques to repair the abdominal wall and prepubic tendon in the postpartum period have been developed, but they are not always successful. In any case, it is unlikely that the mare can be used again for breeding.

PRETERM TORSION OF THE UTERUS

Torsion is rotation of the uterus about its long axis. It is an infrequent but serious complication that occurs with equal frequency in late pregnancy and during foaling. When the uterus rotates 180 degrees (a partial rotation), the arteries and veins that supply the uterus become stretched. This interferes with the blood supply and can devitalize the wall of the uterus. The uterus may rupture and displace the fetus into the abdominal cavity.

A 360-degree complete rotation cuts off the blood supply and results in death of the fetus within a matter of hours. The mare goes into shock and collapses. Complete rotation is rare.

Although the cause of torsion is unknown, rolling of the mare and an exceptionally active fetus have been implicated in some cases.

Signs of partial rotation include mild intermittent colic, frequent urination, listlessness, peering at the flank, kicking at the flank and rolling. These signs may be mistaken for early labor. Any mare who suffers an episode of colic in the last trimester of pregnancy (7.5 months to term) may be suffering from torsion of the uterus and should be seen by a veterinarian at once. Rectal palpation confirms the diagnosis and usually indicates the direction of the twist.

Treatment: When torsion occurs in the preterm mare, the uterus can sometimes be rotated back into position by anesthetizing the mare, placing her on her side, and rolling her in the direction of the torsion. A plank over the abdomen has traditionally been used to stabilize the uterus. This maneuver, however, should not be done near term because of the risk of abortion and rupture of the uterus.

In most cases it will be necessary for your veterinarian to manually reduce the torsion through a flank incision with the mare sedated and standing in stocks. The operator inserts his or her hands through the incision into the abdomen, grasps the uterus, and then rotates it back into position. If the fetus is mature, the next step is to induce labor (see chapter 13).

Treatment of uterine torsion during foaling is discussed in TREATING SPECIFIC DYSTOCIAS in chapter 14.

PRETERM RUPTURE OF THE UTERUS

Preterm rupture of the uterus is usually caused by hydrops of the fetal membranes or torsion of the uterus (both discussed above). When uterine rupture occurs in the preterm mare, signs may be surprisingly mild. Blood loss is often minimal. The foal slips out into the abdomen, and the empty uterus contracts down rapidly.

The diagnosis is made by rectal palpation done for signs and symptoms like those described above for partial torsion. Treatment involves cesarean section followed by repair of the uterus. The operation is done through a midline abdominal incision with the mare anesthetized and lying on her back.

Rupture of the uterus during labor is discussed in POSTPARTUM COMPLICATIONS, chapter 13.

Chapter 8

MARE
INFERTILITY

Infertility can be defined as the inability to produce a live foal. In the broad sense it encompasses failure to breed, failure to achieve pregnancy after breeding, early loss of the embryo, abortion in mid to late pregnancy, and delivery of a stillborn foal.

The major cause of mare infertility is uterine infection caused by bacterial invasion of the reproductive tract *as a result of a breakdown in natural defenses*. Abnormal estrous cycles are the second most common cause of mare infertility (see chapter 2). Silent heat is a behavior problem that can result in infertility if not recognized and treated (see chapter 4, "Determining When to Breed").

THE ROLE OF NATURAL DEFENSES

All mares experience some degree of contamination and infection of the uterus during breeding and foaling. Contamination also occurs during vaginal examinations and the introduction of instruments into the uterine cavity.

The reason that all mares do not remain infected and chronically infertile is that mares with healthy reproductive tracts and natural defenses rid themselves of bacteria, inflammatory debris and infected secretions in a relatively short time.

The endometrium secretes antibodies and immune substances that actively oppose bacterial growth. Estrogen, secreted in its highest concentration at the time of breeding, increases the tone and contractility of the uterine muscle, enabling the uterus to efficiently expel bacteria and infected secretions. This mechanical flushing is considered vitally important in cleansing the uterus and preparing the endometrium to receive the *conceptus*.

The embryo arrives in the uterus 5 days after fertilization. The uterus must eliminate all embryo-toxic products and restore the endometrium to a receptive state of health in less than 5 days. Evidence suggests that mares who are successful breeders are able to clear infected secretions from the uterus in less than 1 day, while those who are unsuccessful can take 4 days or longer.

In addition to immune and hormonal defenses, there are three physical barriers that protect the uterus from becoming contaminated and infected. The first is the seal at the vaginal entrance formed by the lips of the vulva. The second is the vulvovaginal fold. This is a wrinkled flap of vaginal mucosa that runs across the floor of the vagina just behind the external opening of the bladder. This barrier prevents urine from running back into the vagina; it is reinforced by a circle of muscle that forms the vulvovaginal sphincter mechanism, which constricts the vaginal canal. The final barrier is the cervix, which in the nonbreeding season is about the same length and diameter as an index finger. The cervix closes tightly during anestrus, diestrus and pregnancy.

PNEUMOVAGINA

When the seal formed by the lips of the vulva is broken, a windsucking or pneumovagina develops. In mares with pneumovagina, air and airborne contaminants are actively pulled into the vagina and in some cases actually enter the uterus. Loss of the vulvar seal and pneumovagina are the result of changes in perineal conformation associated with aging and the stretching and tearing of the vaginal vestibule that occur with foaling. In addition, a perineal conformation susceptible to pneumovagina tends to be inherited and is much more common in racing breeds.

In a mare with good perineal conformation, the lips of the vulva are nearly vertical, with little or no tilt to the vulvar slit. The lips of the vulva form an effective seal by meeting side-to-side. The top of the vulvar cleft should be at a level with (or only slightly above) the pelvic brim.

As conformation changes with age and the other conditions mentioned above, the vulvar cleft assumes a tilt that exposes the vaginal vestibule to manure from the anus. In addition, a tilted vulva no longer presents an effective seal. The altered anatomy is responsible for the windsucking effect.

A test for windsucking involves separating the lips of the vulva and listening for the characteristic sound of air rushing into the vagina. The test can also be performed by applying pressure with one hand on each side of the vulva.

Another process that occurs with aging and foaling is relaxation of the ligaments supporting the bladder, uterus, vaginal floor and rectum. As these structures drop down into the abdomen, vaginal secretions can puddle around and contaminate the cervix. In addition, the associated sinking inward of the anus allows manure to slide over the open vulva as the mare defecates.

Another condition that accompanies relaxation of the pelvic organs is *urovagina*. During urination, there is no longer a vulvovaginal sphincter mechanism to prevent urine from refluxing into the most dependent portion of the vagina. Urine pooling is a problem particularly during estrus, when the cervix is open and urine can reflux into the uterus.

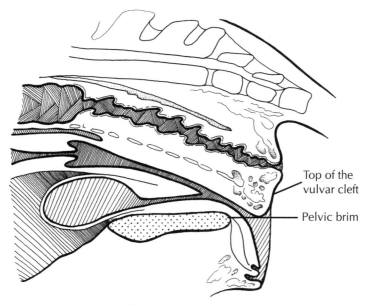

The normal perineum. The vulvar lips are in a vertical position. The top of the cleft is only slightly above the pelvic brim.

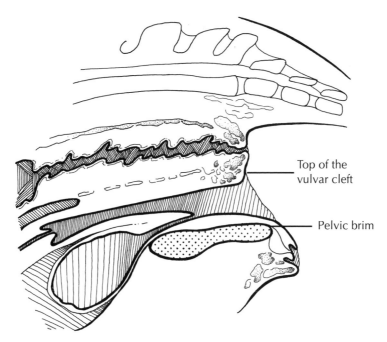

A tilted vulva. Note the angulation of the vulva and the position of the top of the vulvar cleft, well above the pelvic brim.

Treatment: Broodmares with a tilted vulva should undergo surgery to reestablish the vulvar seal and prevent further vaginal contamination. The operation of choice for this purpose is the *Caslick's vulvoplasty.*

This operation is easy to perform and is successful in the majority of mares. It can be done under local anesthesia with the mare sedated and restrained. The procedure involves closing the upper two-thirds of the vulvar cleft. The vulvar lips are infiltrated with local anesthetic. A strip of skin and mucous membrane is removed. The edges of the labia are sutured together leaving an opening below. A breeder's stitch is usually taken just above the lower limit of the closure to reinforce the repair and prevent it from splitting open during breeding. This stitch is a large-diameter suture taken as a single loop between the two sides.

An opening is left just large enough to allow the stallion to enter the mare during breeding. During hand breeding, it may be necessary to manually elevate this opening and guide the stallion's penis through the vulva. A breeding roll may facilitate this process.

If the vulvar passage is too small relative to the size of the stallion's penis, the passage may tear open during breeding, or prevent intromission, entrap the penis, or lacerate the penis on the sharp edges of the breeding stitch. A small opening should be enlarged 2 weeks before breeding. Note also that a vulvoplasty must be opened prior to foaling. A Caslick's vulvophasty that has been opened for breeding or foaling should be closed shortly thereafter.

A *vaginoplasty* is a slightly more complex operation used to repair the vaginal floor after lacerations or stretches associated with foaling. When an obvious inrush of air is heard as the labia are parted, the vulvovaginal sphincter mechanism is no longer intact and should be strengthened by creating a new flap of mucosa. This operation is sometimes combined with a Caslick's procedure.

A *perineal body transection* is the operation of choice for mares with a recessed anus and a sunken vagina associated with urine pooling. It also is the next step when the Caslick's procedure is no longer useable because of the scarring that follows multiple operations to open and close the vulva. This operation involves making a deep incision and spreading the tissues between the anus and vulva. This allows a tilted vulva to drop back into a more normal position.

MARE INFERTILITY EVALUATION

This involves a comprehensive history and physical examination like that described in BREEDING SOUNDNESS EXAMINATION OF THE MARE (chapter 1).

A transrectal ultrasound provides information that cannot be obtained on physical examination. This information may lead to a specific diagnosis.

Hormone tests are useful in diagnosing pituitary and ovarian diseases associated with anestrus and abnormal estrous cycles. Chromosome studies are helpful only in situations where there is an unexplained lack of ovarian function or when the sex of the horse is in question.

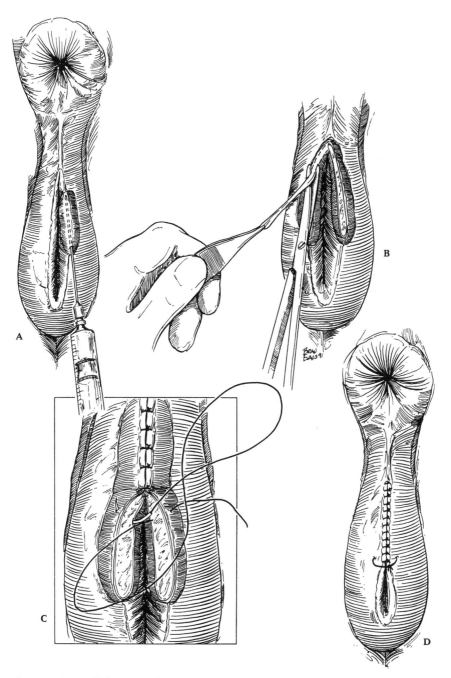

The steps in Caslick's vulvoplasty. Note (D) placement of the "breeder's stitch."

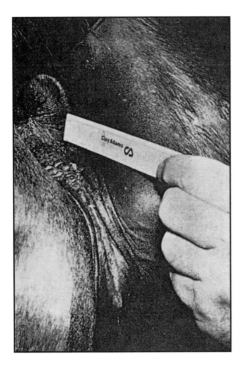

A mare with a sunken anus and tilted vulva that could benefit from a perineal body transection. Courtesy *Equine Reproduction,* by McKinnon and Voss, Lea & Febiger (1993).

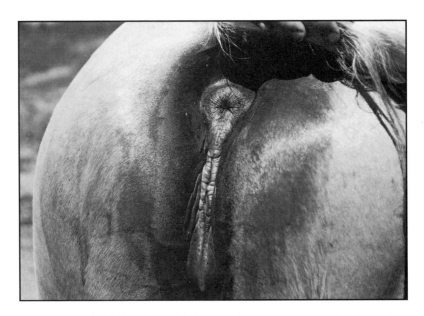

Extreme scarring following multiple operations to open and close the vulva.

AN INFERTILITY EVALUATION IS INDICATED IN MARES WHO . . .

- Fail to conceive after two or more services.
- Incur an early fetal death or abortion.
- Exhibit abnormal estrous cycles.
- Show abnormalities of the reproductive system.
- Remain barren at the end of the breeding season.

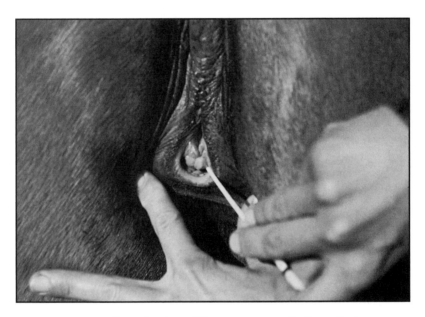

Cultures are taken from the clitoral sinuses to screen for bacteria that cause equine venereal infections.

EQUINE VENEREAL DISEASE

Sexually transmitted venereal diseases can be caused by bacteria, viruses and protozoans.

COMMON BACTERIAL INFECTIONS

Mares who acquire a bacterial infection during breeding develop endometritis and generally do not conceive. If they do become pregnant, they frequently lose the conceptus early in pregnancy. Occasionally a mare with endometritis aborts later in pregnancy or delivers a *septicemic* foal who dies shortly after birth.

Following an acute venereal infection, many mares become asymptomatic and harbor bacteria in the vaginal vestibule, clitoral fossa and sinuses. These asymptomatic mares can pass the infection to stallions.

Stallions who harbor disease-causing bacteria rarely show signs of infection. The bacteria live on the skin of the penis and prepuce, in the urethral fossa, and may be found in the ejaculate. Their presence usually does not affect the stallion's fertility unless the accessory sex glands are involved.

The three bacteria of major importance are *Klebsiella pneumoniae*, *Pseudomonas aeroginosa*, and *Taylor equigenitalis* (the cause of CEM discussed below). These species have all been implicated in outbreaks of venereal infection. They can also be transmitted through contaminated equipment and supplies.

Other bacteria that account for sporadic infections include *Streptococcus zooepidemicus* and *E. coli*. Their critical role in causing mare infection through venereal transmission is questionable. These bacteria are normally present in the reproductive tracts of mares and stallions. It is likely that mares develop infection because of a breakdown in local defenses that protect the uterus.

A delicate balance exists between harmless and disease-producing bacteria on the skin of the stallion's penis. When the protective population of normal flora living in the smegma is disturbed, disease-causing bacteria increase in number and eventually become the dominant species. This increases the potential for causing infection in the mares they service.

Disturbances in the flora of the skin of the penis can be caused by an excessive buildup of smegma or by the excessive and repeated removal of smegma. While it is important to keep the penis clean, excessive washing and the use of detergents and antiseptics such as Betadine or chlorhexidine inflame the skin and remove protective bacteria.

Treatment and Prevention: Mares are treated as described in ENDOMETRITIS below. Stallions are treated as described in ACCESSORY SEX GLAND INFECTION in chapter 10. A period of sexual rest for both sexes lasting at least 2 months is required to prevent relapse. It is essential to reculture the mare or stallion to be sure that the disease has been eliminated before allowing the horse to return to breeding.

Failure to employ a strict program of hygiene for breeding contributes to the transmission of bacterial infections. In addition to washing the mare's vulva and buttocks before mating, it is essential to wash the stallion's penis and sheath both before and *after* mating. Good technique involves wearing disposable gloves. Change gloves between horses. Each horse should have its own separate equipment, including a washbucket with a disposable plastic liner. Rinse the penis repeatedly with warm water (no soap or antiseptic) and avoid vigorous scrubbing. Use clean linen or paper towels and discard after use.

CONTAGIOUS EQUINE METRITIS (CEM)

CEM is a highly contagious venereal infection caused by *Taylorella equigenitalis*. No cases have been reported in the United States or Canada since 1983, but the disease is important to breeders who import horses from countries where CEM is still a problem. These countries include Great Britain, most of continental Europe, and Japan. Horses imported from these areas are subject to quarantine and other restrictions.

Signs in mares appear 6 to 10 days after breeding to an infected or carrier stallion. A copious gray to creamy vulvar discharge mats the hair of the buttocks and tail. In many cases, however, the discharge is absent and the infection is not apparent.

Most mares recover spontaneously, although a significant number become carriers. Infected mares (including those with inapparent infection) are usually infertile during the acute illness. The infertility, however, lasts only a few weeks, after which pregnancy is possible.

Stallions who harbor *Taylor equigenitalis* do not show signs of infection. The first indication of the carrier state is the appearance of CEM and/or lack of pregnancy in the mares serviced by the stallion.

The diagnosis of CEM is made by taking cultures from all accessible sites. In mares, this includes the endometrium, cervix, clitoral fossa and sinuses. In stallions, cultures are taken from the skin folds of the prepuce, urethral fossa, urethra, and the preejaculatory fluid. All samples should be placed in antibiotic-free Amie's or Stuart's transport medium, refrigerated, and transported to an approved testing laboratory within 24 hours of collection.

In mares (only), blood tests are available for detecting antibodies to *Taylor equigenitalis*. These tests become positive 10 or more days after infection. If positive, they indicate only that the mare has had the disease in the past. They do not indicate whether she is a carrier now. There is no blood test for detecting either the presence of infection or the carrier state in stallions.

Treatment: *Taylor equigenitalis* is susceptible to most antibiotics, although the carrier state in mares is difficult to eliminate. Most mares with acute endometritis recover spontaneously. The decision to treat is based on the belief that aggressive treatment may possibly prevent the mare from becoming a carrier.

Recommended therapy involves infusing the uterus with an antibiotic such as penicillin, cleansing the clitoral area with 4 percent chlorhexidine solution and then applying chlorhexidine or nitrofurazone ointment to the clitoral fossa and sinuses. The entire treatment is repeated daily for 5 days.

If a chronic carrier state develops, an operation to remove the clitoral sinus eliminates a continuing focus of infection. Evidence suggests that only the median sinus is involved with *Taylor equigenitalis*. The two lateral sinuses are considered to be too shallow to support bacterial growth.

It is relatively easy to eliminate the carrier state in stallions using local disinfectants. With the penis dropped and the glans extended from the foreskin, the shaft of the penis, including the folds of the prepuce and the urethral fossa, should be cleansed daily for 5 days with a 4 percent chlorhexidine solution. After drying, nitrofurazone cream is applied to these areas.

COITAL EXANTHEMA

This mild venereal disease is caused by equine herpes virus type 3 (EHV-3). The virus can also be acquired during vaginal or rectal examination by contaminated supplies and instruments. Coital exanthema is uncommon in the United States. It does not cause infertility or abortion.

Signs develop 4 to 7 days after sexual contact. Small, painful blisters (less than 1 cm) appear suddenly on the surface of the vulva and perineum of mares. In stallions, blisters develop on the prepuce, glans and shaft of the penis. These blisters develop into pustules that later ulcerate and scab over. Healing occurs in 14 to 21 days.

Ulcerations can become secondarily infected by bacteria. A thick, whitish, mucopurulent discharge develops and the ulcers enlarge and coalesce. The genitals may be inflamed and swollen. Healing is often followed by loss of skin pigment in the ulcerated areas.

Treatment: Breeding should be halted for at least 3 weeks to prevent further transmission. Once the skin is healed, the horse is no longer contagious and it is safe to resume breeding.

Topical antibiotic ointment applied to blisters and ulcers for several days helps to prevent bacterial infection. It is important to keep insects away from open sores.

EQUINE VIRAL ARTERITIS (EVA)

EVA is a contagious respiratory illness spread primarily through contact with infected respiratory secretions. It is discussed here because the virus can also be transmitted in the semen of carrier stallions.

Acute respiratory infection is typified by fever, nasal discharge, conjunctivitis, swelling of the legs, cough and respiratory signs. During the acute illness, the virus enters the bloodstream and invades the placenta, resulting in fetal death and abortion within 1 to 4 weeks.

In the past, EVA was responsible for abortion storms involving 50 to 80 percent of pregnant mares on a single farm. The last epidemic occurred in central Kentucky in 1984. Since that time, the disease has occurred rarely, although sporadic cases have been reported in California, Arizona, Colorado, Indiana, Kentucky, New York, Ohio and Pennsylvania.

Serologic tests are available to determine whether a horse has ever been exposed to EVA. The widespread prevalence of seropositivity among certain breeds suggests that many of these horses acquire asymptomatic infection. Standardbred and American Saddlebred horses, for example, have a high incidence of seropositivity (85 and 25 percent, respectively), while Thoroughbreds have a low incidence (2 percent).

Stallions can develop a carrier state and shed the virus in their semen for months or years. Mares do not become carriers. However, if a seronegative mare is bred to a carrier stallion, she can come down with a respiratory infection and pass it on to other horses, including pregnant mares.

Serologic blood tests become positive 2 weeks after infection. A rising-antibody concentration based on paired samples taken 2 weeks apart indicates recent exposure with active antibody formation. A steady level can be the result of either infection or vaccination in the past. If a stallion is seropositive but was never vaccinated, his antibodies must have been acquired as the result of exposure. He may or may not be a carrier. Of seropositive stallions, 30 to 50 percent are carriers.

To determine if a stallion is a carrier, the virus can be isolated from the sperm-rich fraction of his semen. The ability to recover viruses is highly dependent on the capabilities of the local laboratory and is not considered absolutely reliable.

The results of test breedings can also be used to determine if a seropositive stallion is a carrier. For example, if a negative mare becomes positive after breeding, the stallion was the source of the infection.

There is no specific treatment for EVA and no way to eliminate the carrier state in stallions.

Prevention: EVA virus can be recovered from respiratory secretions for up to 3 weeks after illness. In breeds with a high incidence of EVA infection, it is a good policy to isolate new arrivals for 3 weeks. This will prevent resident horses from acquiring the disease from newly arrived horses.

Stallions known to be carriers should be bred only to seropositive mares. A mare can become seropositive either through exposure or vaccination. An effective vaccine is available. Adequate protection is afforded by a single immunization and annual boosters. The first vaccination must be given at least 21 days before breeding. Do not vaccinate pregnant mares.

EVA vaccine is controlled by state regulatory officials and is issued in accordance with established guidelines that vary from state to state. States in which the disease is endemic usually follow the guidelines developed by the Commonwealth of Kentucky. Regulatory information on EVA vaccine, can be obtained by contacting the State Veterinarian's office.

DOURINE

Dourine is a serious *systemic* infection caused by protozoans transmitted during breeding. Dourine has been eradicated in Europe and the United States but still occurs in the Middle East, North and South Africa, Central and South America, and parts of Asia and the former USSR.

The incubation period is 1 to 2 weeks. The onset is slow. The initial sign is a mild, recurrent fever. Swelling of the vulva and a mucopurulent discharge occur in mares. Stallions have a urethral discharge accompanied by swelling of the penis, prepuce and scrotum. As the disease progresses, large, round patches of raised skin appear in crops all over the body. The horse becomes emaciated and develops hindquarter paralysis.

Dourine is often fatal. Treatment is impractical. Euthanasia is mandatory to prevent epidemic spread. State and federal authorities should be contacted if the disease is suspected.

DISEASES OF THE UTERUS

ENDOMETRITIS

Endometritis is an infection of the lining of the uterine cavity. It is the most common cause of infertility in mares. Poor perineal conformation, uterine contamination during breeding and foaling, loss of uterine defense mechanisms associated with multiple pregnancies, and equine venereal diseases are all predisposing causes. The older, multiparous mare, who is no longer able to clear her uterus of infective bacteria, is the classical example of a mare most likely to be suffering from endometritis.

Endometritis is classified as *acute* or *chronic,* depending on the duration of the infection. Bacterial species most frequently isolated from the endometrium, in order of frequency, are *Streptococcus zooepidemicus, E. coli, Pseudomonas aeroginosa* and *Klebsiella pneumoniae.*

A mare with endometritis appears healthy but either fails to conceive after repeat breedings, or conceives and loses the pregnancy in early gestation (see EARLY EMBRYONIC DEATH, chapter 9). A vaginal discharge is occasionally seen, but is more likely to be detected on vaginal speculum exam. Rectal palpation is often normal but may reveal an enlarged uterus containing fluid. This can be confirmed by ultrasonography.

The diagnosis is made by finding inflammatory cells in swabs taken from the uterine cavity. Endometrial cultures aid in choosing effective antibiotics. An endometrial biopsy provides information about the degree of endometrial scarring. The above samples are best taken during mid-estrus, when the cervix is open and uterine defense mechanisms are most active.

Endometrial Cytology. Cytology is the key to the diagnosis of endometritis. By definition, endometritis is an inflammatory process and therefore should be confirmed by finding inflammatory cells. Inflammatory cells are polymorphonuclear leukocytes (PMNs), the same as white blood cells. Together with tissue-breakdown products, the white cells make up pus. If pus is identified in the uterine cavity or is seen coming from the cervix, a diagnosis of infection is confirmed. In many cases of chronic endometritis, however, PMNs are present in small numbers and must be identified microscopically.

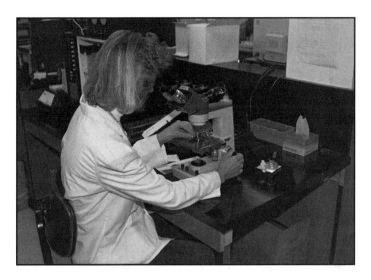

The key to diagnosing endometritis is finding inflammatory cells on a cytology specimen taken from the mare's endometrium.

A cytology specimen can be obtained with a culture swab as described below for obtaining endometrial cultures. The tip of the swab is rolled back and forth on a clean glass slide and fixed with a commercial spray. In the laboratory, the slide is stained and examined under the microscope for PMNs, bacteria and fungi.

A second method of collecting endometrial cells involves flushing the uterus with large volumes of sterile saline. A cloudy return indicates pus. When the return is clear, the specimen is spun down and the sediment is examined microscopically.

A mare with a healthy endometrium has few if any PMNs and no bacteria in a high-power microscopic field. A mare of indeterminate status will have several PMNs and perhaps an occasional bacterium in the microscopic field. An obviously infected mare will show numerous PMNs interspersed with many bacteria. Indeterminate findings must be compared with the results of endometrial biopsies and cultures.

Endometrial Culture. There are two methods for obtaining cultures of the endometrium. Both seek to prevent sample contamination by keeping the swab from contacting the cervical or vaginal secretions.

In the unguarded method, a sterile tubular vaginal speculum is passed into the vagina. A long, sterile swab is passed through the speculum and into the uterine cavity. The swab is rotated to ensure maximum contact with the surface of the endometrium. It then is withdrawn through the speculum and inoculated onto a transport medium in a glass tube.

The guarded method (which is the preferred method) uses a sterile swab inside a plastic case. The case is guided with a lubricated, gloved hand through the cervix and into the uterus. Once in the uterus, the swab is extended, the sample obtained, and the swab retracted back into its case. The plastic case then is withdrawn. Cultures obtained with the guarded method cannot be contaminated with vaginal or cervical secretions.

A heavy bacterial growth on an agar plate strongly suggests active infection.

In the laboratory, the culture specimen is inoculated onto agar plates. The plates are examined for bacterial growth during the next 2 to 3 days. A report of "no growth" indicates a sterile culture. "Mixed growth" is of questionable significance. It usually indicates contamination but may indicate that bacteria are minimally present in the uterus. "Pure growth" of known disease-causing bacteria is highly significant.

Endometrial Biopsy. An endometrial biopsy provides additional information, particularly whether chronic infection has resulted in fibrosis (scarring). Widespread scarring of the endometrium interferes with the ability of the mare to conceive and maintain a pregnancy.

A biopsy can be done at any stage in the estrous cycle. The biopsy instrument is guided through the cervix, and a bite of endometrium is taken. The specimen is placed in 10 percent formalin and sent to the laboratory.

Findings on microscopic examination include (a) normal endometrium for the phase of the estrous cycle, (b) acute inflammatory cells (PMNs), (c) chronic inflammatory cells (plasma cells and lymphocytes), and (d) atrophy of glandular tissue with replacement by fibrous connective tissue.

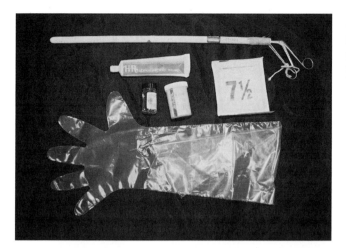

Equipment needed for endometrial biopsy.

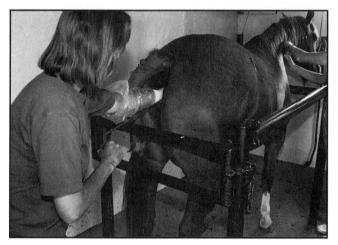

The biopsy forceps is guided through the cervix. A bite of tissue is taken from the endometrium.

The biopsy is placed in formalin.

The presence of PMNs on biopsy, along with growth of *pathogenic* bacteria on culture, confirms the diagnosis of acute endometritis.

A chronically infected endometrium will often show a wasting away of glandular cells, along with a buildup of scar tissue around the mucous glands. The magnitude of these changes can be used to predict whether a mare is likely to carry a foal to term. Mares in Category Grade 1 have a normal endometrium and excellent foaling rates. Mares in Category Grades 2(a) and 2(b) have mild to moderate endometrial scarring, respectively, with foaling rates in the neighborhood of 50 percent. Mares in Category Grade 3 have widespread inflammation and fibrosis with glandular atrophy. Foaling rates are in the neighborhood of 20 percent. If these mares do become pregnant, they often develop placental insufficiency and subsequently abort (see chapter 9).

If scarring is severe enough to obliterate the entire endometrium, the likelihood of pregnancy is nil.

Treatment: The objective is to eliminate infection or at least reduce inflammation to the point at which fertility will not be compromised. Mares with *acute* endometritis and purulent vaginal discharge should be treated when the diagnosis is made. Treatment is described below.

Mares with chronic inflammation can be treated during estrus or immediately after breeding. There are two approaches. One relies on the use of antibiotics. To be truly effective, the antibiotics should be *bactericidal* against the species of bacteria in question. This is determined by sensitivity tests. Antibiotics are administered directly into the uterus. The advantage of local infusion is that it produces a high concentration of drug at the site of infection. The disadvantage is that it facilitates the emergence of antibiotic-resistant bacteria. This is becoming a common problem.

The second approach, preferred by many veterinarians, involves merely infusing the uterus with 1 to 2 liters of sterile saline. Flushing out the uterus greatly reduces the amount of fluid, debris and harmful by-products of infection trapped in the uterine cavity, allowing the mare's natural defenses to cope with the rest. Uterine irrigation does have the potential for introducing new bacteria into the uterus. However, it does not result in the development of antibiotic-resistant strains. The procedure is usually carried out during estrus. At this time, the cervix is open and the endometrium is under the protective influence of estrogen.

The procedure for uterine irrigation involves passing a long, rubber catheter with a deflated cuff through the cervix into the uterine cavity. The cuff is inflated to seat the catheter. A plastic pipette can be used in place of a catheter. The uterus is infused with the sterile saline through the catheter or pipette. The fluid is allowed to return by gravity. The appearance of the recovered fluid is noted. Cloudy fluid indicates the need to repeat the irrigation. Clear fluid indicates that the uterus has been adequately cleansed.

In the treatment of *acute* endometritis, uterine irrigation is often combined with local antibiotics. In one protocol, antibiotics are infused every other day for a total of five infusions. On days when antibiotics are not infused, the uterus is flushed with sterile saline alone.

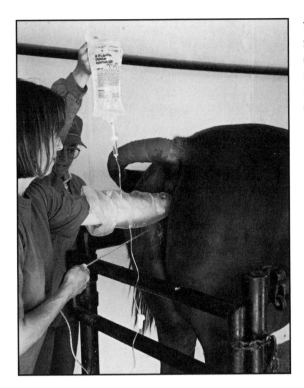

Treatment of endometritis. Saline irrigation of the uterus washes out pus and inflammatory debris. Natural defenses deal with the residual inflammation.

It is relatively common for chronic endometritis to flare up immediately after breeding. Uterine infusion may reduce the level of inflammation and increase the chance of pregnancy. The uterus is flushed just prior to breeding and then again 6 hours after breeding. The flushing can be repeated daily until the 3rd day after ovulation.

There is little possibility that flushing will interfere with conception or cause premature regression of the corpus luteum. The sperm are out of the uterus and into the tubes by 2 hours after mating. The infusion is unable to enter the tubes because of the presence of valves. The corpus luteum cannot respond to prostaglandin released by the endometrium in response to a uterine manipulation before the 5th day postovulation. Finally, the fertilized egg does not enter the uterus before the 5th day postconception.

One other method of treating chronic endometritis involves increasing the frequency of estrus by shortening the estrous cycle (see SHORTENING THE ESTROUS CYCLE in chapter 2). Shortening the estrous cycle extends the influence of estrogen on the endometrium. An estrogen-stimulated endometrium is more effective in eliminating inflammation.

Prevention: Attention to good hygienic practices during breeding and foaling will reduce the extent and occurrence of uterine contamination. The mare and stallion should be washed before breeding, as described in chapter 5, "Breeding." A further reduction in exposure can be accomplished by limiting the number of services through monitoring of estrus and ovulation, and by using artificial insemination when permitted by the breed registry.

It is important to correct any anatomical problems, such as pneumovagina or incompetent cervix. Invasive diagnostic procedures should be performed with antiseptic precautions, and only when medically necessary.

PYOMETRA (ABSCESS OF THE UTERUS)

The accumulation of pus in the uterus is called pyometra. It differs in many respects from endometritis. The problem in pyometra is that the uterus is unable to evacuate infected material which accumulates and forms a large abscess in the uterine cavity.

Some cases are caused by a blocked cervix resulting from scars and adhesions that follow cervical tears. In most cases, however, the cervix is normal. These mares appear to have a functional problem in that the cervix fails to open completely during estrus.

An abscessed uterus can hold up to 60 liters of pus, although this amount is unusual. Intermittent vaginal discharge occurs when pus escapes through the cervix.

Mares with pyometra do not appear ill and seldom develop fever and other signs of toxicity. Estrous periods are often irregular due to a persistent corpus luteum (see chapter 2). Lack of heat activity is often the reason the veterinarian is asked to examine the mare.

Rectal palpation discloses an enlarged uterus, which can also be found incidentally on a routine prebreeding exam. A transrectal ultrasound reveals a uterus filled with fluid. This confirms the diagnosis.

Treatment: Treatment involves placing a catheter through the cervix into the uterine cavity. The pus is evacuated and the cavity flushed with saline until the fluid return is clear. Antibiotics may be instilled at the conclusion of the lavage. The mare should be treated for cervical adhesions and stenosis (if present), as described elsewhere in this chapter. Pyometra is often followed by chronic endometritis; further treatment is directed at that problem. An endometrial biopsy can be used to determine the extent of scarring and mucous gland atrophy. This gives a prognosis for future fertility.

Mares with persistent or recurrent pyometra may continue to have intermittent vaginal discharge. The decision to re-treat depends on the mare's breeding potential and/or whether the vaginal discharge interferes with her athletic performance or recreational use. Hysterectomy (removal of the uterus) can be considered in some cases.

CERVIX DISEASES

CERVICITIS

Infection of the cervix is nearly always part of a more extensive process involving the vagina or uterus. Uncommonly, pregnant mares can develop an isolated infection of the cervix.

Vaginal discharge is the only sign of cervicitis. The discharge often contains mucus produced by glands inside the cervical canal.

Treatment: The cervix should be cultured and the mare treated with a course of antibiotics selected on the basis of sensitivity tests. The mare should be investigated for an associated endometritis. If present, treatment is directed at that problem.

Cervical infection is sometimes complicated by scarring and the development of stenosis (discussed below).

LACERATIONS AND CERVICAL INCOMPETENCE

Lacerations of the cervix occur during foaling and rarely during breeding. Most tears involve the rim of the cervix and extend into the cervical canal. A deep tear divides the sphincter-like muscle of the cervix.

Vaginal bleeding after foaling suggests a lacerated cervix. However, it is difficult to see the laceration at this time because the cervix is enlarged and swollen. The best way to determine if a laceration exists is by feeling the wall of the cervix between the fingers and thumb.

Most lacerations are not detected at the time of injury. The torn or scarred cervix is discovered on digital palpation during a routine prebreeding examination or an infertility workup.

Lacerations that heal with the edges apart (in the open position) may result in cervical incompetence. An incompetent cervix is one that is unable to close completely during diestrus and pregnancy. The partially open cervix becomes a portal for bacteria to enter and infect the endometrium or placenta.

An incompetent cervix is best evaluated by inserting a finger into the cervix and palpating the entire inner surface of the canal. The examination should be performed during diestrus, when the cervix is contracted under the influence of progesterone. A cervix that is not completely closed in diestrus is incompetent.

Treatment: Lacerations involving the muscular wall of the cervix should be repaired to preserve breeding potential. If the injury is detected at foaling, repair is best delayed 1 month to allow for complete healing. An incompetent cervix found on routine examination should be repaired prior to breeding. This must be done in diestrus or anestrus. Surgery can be performed under epidural anesthesia with the mare in stocks. Special vaginal retractors and extra-long instruments are needed.

An incompetent cervix in a pregnant mare leads to abortion. Treatment of an incompetent cervix discovered during pregnancy is discussed in chapter 9.

CERVICAL ADHESIONS AND STENOSIS

Cervical adhesions are bands or sheets of fibrous connective tissue that develop over the cervix and within the canal. These adhesions can be caused by vaginitis, cervicitis, cervical lacerations and the introduction of irritant solutions into the mare's reproductive tract.

Stenosis is a narrowing of the cervical canal caused by scarring and contracture of the sphincter muscle. Both cervical adhesions and stenosis interfere with the opening and closing of the cervix during estrus and diestrus. As a result, these conditions are associated with endometritis, pyometra and incompetent cervix.

Treatment: Adhesions can be divided with long scissors or broken up with fingers and instruments. Stenosis is treated by dilating (enlarging) the cervical canal. Because scar tissue grows back rapidly, the treatment may need to be repeated several times. Application of a steroid-antibiotic ointment helps to slow the rate of recurrence. Placing a catheter through the cervix may keep the cervix open as adhesions reform.

VAGINAL AND VULVAR DISEASES

VAGINITIS

Infection of the vulva and vagina occurs after breeding, foaling, and the introduction of contaminants into the vagina. A mare with a uterine infection may have a secondary vaginitis. *Pneumovagina* is a major cause of vaginitis.

Urovagina occurs in mares with a defective *vulvovaginal sphincter* mechanism. Residual urine in the vagina invariably results in secondary vaginal infection.

The principal sign of vaginitis is a mucopurulent discharge in a mare who otherwise appears healthy. On vaginal speculum exam, the mucosa appears inflamed and swollen.

Vaginitis and vulvitis are mild infections that do not cause infertility. The importance of vaginitis is that it can progress to endometritis, particularly if there is a continuing source of contamination.

Uncomplicated vaginitis usually clears spontaneously. Vaginal washes with a dilute antiseptic such as 2 percent chlorhexidine may hasten this process. Treatment of inflamed skin (urine scalds) associated with urovagina involves the periodic application of a zinc oxide ointment. Perineal deformities that predispose mares to vaginal contamination must be surgically corrected to preserve breeding potential.

VAGINAL BLEEDING

A maiden mare with an intact hymen will experience slight vaginal bleeding after breeding.

A mare who bleeds after covering may have sustained a vaginal tear or a laceration of the cervix. Such injuries are most likely to happen when a small mare is bred to a stallion with a large penis. When blood is seen on the mare's vulva or the stallion's penis after breeding, veterinary examination is indicated.

Breeding injuries can be minimized by knowledgeable stallion handling and by using a breeding roll in selected mares that prevents complete entry of the penis. An intact hymen should be opened before the breeding season.

Postpartum vaginal bleeding is discussed in chapter 13, "Foaling."

RECTOVAGINAL FISTULA

A rectovaginal fistula is an abnormal connection between the rectum and vagina. The opening from the rectum into the vagina may be close to the perineum or located within the *vaginal vestibule*.

A fistula develops when a tear in the roof of the vagina extends through the perineal body into the rectum. In almost all cases, this injury occurs during the second stage of labor when the foal's foot punches through the vagina and then is rapidly withdrawn.

A rectovaginal fistula allows fecal material to contaminate the mare's entire genital tract. The mare is incapable of fertility until the fistula is repaired.

Treatment: It involves surgical closure of the fistula. If the mare has a foal at her side, the operation should be postponed until the foal is weaned, even though it may mean skipping the rest of the breeding season. Surgical repair is complicated and often must be done in two stages. Postoperatively, keep the manure soft by feeding pasture and bran. It may be necessary to administer laxatives and stool softeners by stomach tube. Enemas are given if the mare develops hard manure and strains to defecate.

VULVAR DISEASES

The tilted vulva associated with *pneumovagina* is the most common vulvar deformity. However, badly healed lacerations and congenital deformities of the vulva can also prevent the lips from meeting and forming an effective seal.

A mare with a tilted vulva may have had a Caslick's operation to close the upper two-thirds of the vulvar cleft. The remaining opening should be large enough to permit successful intromission and breeding. If there is any question about the adequacy of the opening, consult with your veterinarian before breeding the mare.

Tumors of the vulva include the melanoma, squamous cell carcinoma and sarcoid. A *biopsy* is the best way to make an exact diagnosis.

Melanomas are firm, hairless nodules that occur beneath the skin, frequently in the perineum and on the vulvar lips. These growths are often multiple. They occur in aged gray or white horses. Melanomas are usually benign, but when present for many years can become malignant and *metastasize*. They generally are not removed unless they become large enough to interfere with urination or defecation. Oral cimetidine (Tagamet™) has been associated with the regression and disappearance of melanomas in some horses.

Squamous cell carcinomas are fleshy growths or flat ulcerations. These tumors rarely metastasize and can be treated successful by surgical removal or cryosurgery (freezing), depending on size and location.

Sarcoids are the most common tumors of horses. They occur everywhere on the body; there is no specific preference for the vulvar area. Some sarcoids are rough and wart-like. Others are raised nodules or red-colored growths. An uncommon type is flat and hairless.

Sarcoids have been found by DNA studies to be caused by the cattle wart virus. They often develop at sites of prior trauma. This suggests that the virus can be introduced through cuts and wounds. Sarcoids do not metastasize but have a high rate of recurrence after surgical removal.

OVARIAN DISEASE

TUMORS OF THE OVARY

Ovarian tumors make up about 5 percent of all *neoplasms* in the horse. Accordingly, they are relatively uncommon.

Benign ovarian tumors include the granulosa-thecal cell tumor, the cystadenoma and teratoma. Teratomas may contain hair, bone, cartilage and teeth. Cystadenomas and teratomas have little effect on the estrous cycle. They usually are discovered incidentally during rectal palpation. The treatment of benign tumors involves removing the diseased ovary.

Dysgerminomas are rare *malignant* tumors that are usually widespread when first discovered. Treatment generally is impossible.

Granulosa-thecal Cell Tumor. The most common tumor of the ovary is the benign granulosa-thecal cell tumor. It usually occurs in mares 5 to 9 years of age. The typical presentation is that of a mare who fails to come into heat during the breeding season. Rectal palpation reveals an enlarged ovary with no palpable ovulation fossa. Tumors the size of a watermelon have been described. The opposite ovary is small, firm and without follicles. An ultrasound of the ovaries reveals the tumor to be a solid mass with a honeycombed appearance.

Granulosa-thecal cell tumors can produce estrogen, progesterone or testosterone, although many do not. Tumors that produce estrogen are responsible for the

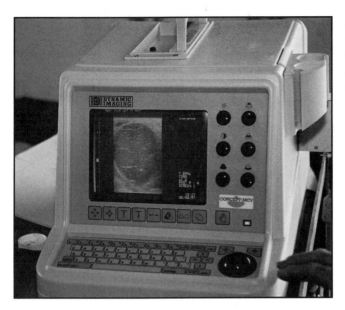

A 7 cm mass in the left ovary that proved to be a granulosa-thecal cell tumor.

continuous estrus behavior seen in a small number of mares. This persistent estrus behavior may be the first indication that the mare has a tumor of the ovary.

Tumors that produce progesterone cause diestrus-like behavior. The mare remains persistently disinterested in the stallion and does not exhibit heat.

Testosterone-secreting tumors cause the mare to behave like a male. These mares often have a crested neck and an enlarged clitoris.

After removal of a granulosa-thecal cell tumor, the opposite ovary usually regains normal cyclic activity and the potential for pregnancy within 2 to 18 months.

CYSTIC OVARY SYNDROME

A cyst is a bubble-like structure that contains fluid. Mares commonly develop cysts around the ovary. These paraovarian cysts are occasionally noted as incidental findings on ultrasound examinations. They have no reproductive significance.

Cystic Ovary Syndrome. On rare occasions during the breeding season a mare may develop a large, persistent ovarian follicle that resembles a simple cyst on ultrasound. These follicles contain estrogen and may be associated with persistent estrus or irregularities of the estrous cycle. The basis of these large follicles is unknown.

CHROMOSOME ABNORMALITIES

Sex chromosome abnormalities are a leading cause of primary ovarian infertility. Most occur spontaneously. They are not inherited, because the affected horse is nearly always sterile. Genetic studies are required to make a diagnosis.

The most common sex-chromosome disorder is *ovarian hypoplasia*, also called *gonodal dysgenesis*. Affected mares have a clitoris, vulva and vagina that appear to be normal. On rectal palpation, the ovaries are small, smooth and firm. The uterus and cervix are small or underdeveloped. These mares may exhibit occasional or very irregular signs of heat but rarely develop ovarian follicles.

Three intersex abnormalities produce various types of hermaphrodism. The *male pseudohermaphrodite* is the most common. This horse is genetically a female but has undescended abdominal testicles. The enlarged clitoris resembles a short penis. The vagina is short and terminates at the urethral opening. Behavior is like that of a stallion, or nymphomaniac mare (discussed below).

The *female pseudohermaphrodite* and the *true hermaphrodite* are exceedingly rare. The female pseudohermaphrodite has ovaries, but the external genitalia are primarily male. The true hermaphrodite has male and female genitalia and remnants of both male and female sex glands.

Removal of the testes is recommended for the masculinized, aggressive mare with male pseudohermaphrodism. The temperament of these horses may improve after neutering.

BEHAVIORAL CAUSES OF INFERTILITY

BEHAVIORAL ANESTRUS (SILENT HEAT)

Behavioral anestrus is a common problem in mares. It is characterized by normal estrous cycles but failure to show heat to the stallion. This is not a true anestrus, since these mares do ovulate and cycle. Treatment is discussed in PROBLEMS WITH DETECTING HEAT (chapter 4).

NYMPHOMANIA

This is a rare condition in which the mare exhibits an exaggerated or wanton display of estrus, yet paradoxically rejects the stallion. The cause is unknown. Nymphomaniac mares have a normal reproductive system and cycle at regular intervals. Estrogen concentrations are within the normal range.

The nymphomaniac mare squirts urine, swishes her tail, squeals when touched about the hindquarters and is extremely aggressive toward horses and people. In the mild form, this behavior is less marked and occurs primarily at the time of heat. In the severe form, the behavior occurs throughout the estrous cycle.

Breeding is difficult or impossible because of the mare's violent aggression toward the stallion. It is hazardous to perform unrestrained rectal examinations on nymphomaniac mares—even those with a mild form of the condition.

The behavior of mares with mild nymphomania may improve with the removal of both ovaries. Severely affected mares are unlikely to respond to any form of treatment. Euthanasia is often recommended.

Nymphomania-like behavior is occasionally caused by an estrogen-producing granulosa-thecal cell tumors of the ovary (discussed above).

STALLIONLIKE BEHAVIOR

It is not normal for mares to mount other horses or exhibit behavior such as vocalization, teasing, herding, urine-marking and unprovoked aggression toward stallions.

When stallionlike behavior does occur, it is usually because the mare is being given (or was given in the past) an anabolic steroid containing testosterone to improve racing performance (not an approved use). These mares frequently exhibit masculine behavior and have clitoral enlargement for months after blood testosterone levels have returned to baseline norms (see chapter 5).

Male-type behavior also can occur with an androgen-producing granulosa-thecal cell tumor of the ovary, discussed above.

Pregnancy is an infrequent cause of masculine behavior. The behavior appears at about 4 months' gestation and continues throughout pregnancy. The behavior coincides with the period during which the placenta produces steroids that are precursors of testosterone. Thus the placenta may be the source of a masculinizing hormone, although the relationship has yet not been proven. The male behavior usually disappears after foaling.

Nonpregnant mares with unexplained masculine behavior should be examined for a tumor of the ovary.

ABORTION

GENERAL INFORMATION

The risk of pregnancy loss may be as high as 30 percent. The danger is greatest during the period of the embryo. Fetal loss is more common in mares over 18 years of age and in mares with a history of prior abortion. Endometritis is the leading cause of fetal loss. It is followed in frequency by twin pregnancy and the rhinopneumonitis virus.

By definition, the *conceptus* remains an embryo until day 40 of gestation. After day 40, it is called a fetus. Pregnancy loss before day 40 is called early embryonic death. Loss of the pregnancy between 40 and 300 days' gestation is called abortion. A foal born after 300 days is potentially capable of independent life outside the womb. Accordingly, if the foal is born alive after 300 days but before 320 days, it is called premature. A stillbirth is any foal born dead after 300 days' gestation.

When a mare aborts, the cause should be determined whenever possible. Diseases that cause abortion may present health hazards to other horses and the general public. Knowing why the mare aborted might make it possible to prevent the same thing from happening in the future.

When abortion occurs after the 2nd month of pregnancy, laboratory examination of the fetus and placenta will reveal the cause in about 50 to 60 percent of cases. Without postmortem examination, the chance of finding out what caused the abortion is less than 10 percent. Diagnostic laboratories prefer to have the complete fetus and placenta submitted in a chilled (not frozen) condition as soon as possible after the abortion. In some cases, shipping a complete fetus is not practical. In these situations, your veterinarian can perform several studies in the field and obtain tissue samples to send to the laboratory.

Fetal age is estimated by the presence or absence of hair on the mane, muzzle, eyebrows and coronary bands. Measurements of the crown-rump length and umbilical cord are compared to standards to determine whether fetal development was appropriate for gestational age. Discrepancies indicate fetal growth retardation or an error in breeding records. The placental membranes are carefully inspected for areas of absent *chorionic villi* and for inflammatory changes indicating placental infection. Blood is obtained from the free end of the umbilical cord.

Visual inspection of the organs of the fetus may provide clues as to the presence or absence of fetal disease, particularly intrauterine sepsis. Specimens are taken from various organs as well as the placenta. The tissue is placed in sterile culture tubes and in glass tubes containing formalin. The samples are packed in ice and transported to a diagnostic laboratory equipped to do *histology* and virus isolation, as well as cultures for bacteria and fungi.

The cause of an abortion can be classified as infectious, noninfectious, or unknown. The noninfectious category is broad. It includes twinning, trauma, genetic defects, and toxicities such as forage poisoning. The infectious category includes fetal infection, maternal infection, and placental infection.

Mares who abort on successive pregnancies should be suspected of having chronic endometritis or twin pregnancies. A less-common cause is progesterone deficiency, discussed below.

Any mare who loses an embryo or fetus should have an infertility evaluation (see chapter 8).

EARLY EMBRYONIC DEATH

Early embryonic death (EED) is loss of the embryo before 40 days' gestation. Recent ultrasound scanning in early pregnancy reveals that EED is common. In fact, the incidence may be as high as 20 percent in fertile mares and greater than 70 percent in subfertile mares. Many embryos are lost during the critical period at 5 to 6 days' gestation when the embryo enters the uterus.

Signs of EED are rarely noted. Unless a pregnancy is detected by an ultrasound scan or rectal palpation and is subsequently found to be absent, the only indication of a missed pregnancy is a failure to return to heat at the expected time. Vaginal bleeding usually is not seen. These early embryos are small and presumably are reabsorbed. However, in one study when EED occurred after implantation (day 16), ultrasound scanning of the cervix revealed that the cervix was open. This suggests that some embryos are lost by passage through the cervix.

When EED occurs after day 36, the endometrial cups (discussed in PSEUDOPREG-NANCY, chapter 2) continue to produce equine chorionic gonadotropin. This prevents the mare from returning to heat until after the cups cease to function at 110 to 120 days' gestation. In effect, breeding may have to be postponed until the next season.

Chronic endometritis is an important cause of early embryonic loss. Many mares with endometritis are capable of pregnancy but incapable of maintaining a pregnancy for one of three reasons. The first is that during the 16 days when the fertilized embryo is unattached, it is nourished by a uterine "milk" secreted by endometrial glands. A damaged endometrium may not be able to produce enough milk to adequately nourish the free-floating embryo. Second, an infected endometrium might release prostaglandin that prematurely terminates the *corpus luteum* (CL) and eliminates the progesterone on which the pregnancy depends. Third, endometritis may be lethal because of direct effects of bacteria and inflammatory products on the embryo.

Progesterone deficiency has long been considered a cause of unexplained EED, although low serum progesterone concentrations in early gestation have not been well-documented in horses. One exception would be the mare with chronic endometritis and premature regression of the CL.

Many chromosomal defects are incompatible with early embryo development. Chromosomal defects, however, are much less common in horses than in humans and should account for only an occasional case of EED.

Prevention of EED: Chronic endometritis can be diagnosed by a prebreeding examination. Endometrial inflammation and glandular atrophy in mares with Category 2(a) and early Category 2(b) changes, described in the section on EN-DOMETRITIS (chapter 8), can often be improved by treating the endometritis and correcting predisposing factors, such as pneumovagina, urovagina and incompetent cervix. The risk of acquiring sexually transmitted diseases can be minimized by good breeding practices, as described under EQUINE VENEREAL DISEASE, chapter 8.

Progesterone therapy has been successful in some cases of recurrent abortion. A specific indication is the mare with chronic endometritis and premature regression of the CL. The progesterone simply replaces what would normally be produced by the mare's ovaries.

To be effective, progesterone must be administered in adequate doses throughout the first 4 months of gestation. The placenta then takes over this function. There are no known harmful side effects of giving progesterone, although clitoral enlargement has been reported in fillies whose dams were treated with altrenogest.

Estrogen deficiency is not a known cause of EED, and estrogen therapy has no benefit in preventing either EED or abortion.

INFECTIOUS CAUSES OF ABORTION

Assume that all abortions are infectious. Use rubber gloves and handle the tissues with sterile precautions.

EQUINE RHINOPNEUMONITIS (HERPES VIRUS 1)

Rhinopneumonitis is a highly contagious viral respiratory infection in horses. It is also a leading cause of infectious abortion in mares. Abortion storms have been reported involving up to 70 percent of pregnant mares on a single farm.

There are two viral subtypes associated with rhinopneumonitis infection. EHV-1 causes nearly all cases of abortion and some cases of respiratory infection. EHV-4 accounts for the majority of respiratory infections but rarely causes abortion.

Immunity to the herpesvirus following natural infection or vaccination persists only for a short time. Repeated infections of susceptible individuals keep the virus endemic in the horse population.

Pregnant mares are usually exposed to rhinopneumonitis during the respiratory season in fall and early winter. Signs of infection include nasal discharge ("snots"), eye discharge and a dry cough. These signs may be absent when infection is caused by EHV-1. The virus enters the bloodstream and invades the placenta and fetus.

Abortion does not occur until late pregnancy (between 7 months and term). Death is caused by premature separation of the placenta, with suffocation of the foal. Occasionally a mare goes to term and delivers a sick foal who dies shortly after birth (see FOAL PNEUMONIA in chapter 15).

Diagnosis. Examination of the fetus and placenta reveals changes characteristic of widespread inflammation. Viral inclusion bodies may be seen in the liver and other tissues. Isolation of the virus is conclusive but not always possible. Antibodies to EHV-1 may be found in fetal blood.

Prevention: Vaccination of pregnant mares is highly recommended, even though 100 percent protection is not always achieved. Owing to the short duration of protection, booster shots should be given at 5, 7 and 9 months' gestation using only killed (inactivated) EHV-1 vaccines (see Tables II & III, appendix). Rhinopneumonitis vaccines containing EHV-4 are available and may be the vaccines of choice for preventing respiratory infection in foals, weanlings, yearlings and pleasure horses. Broodmares, however, should continue to be vaccinated with EHV-1, since this is the subtype responsible for most abortions.

EQUINE VIRAL ARTERITIS (EVA)

EVA is a contagious viral illness associated with placental infection and abortion 1 to 4 weeks after illness. The disease is transmitted via the respiratory route, and also by viruses shed in the semen of carrier stallions. EVA is discussed in detail in EQUINE VENEREAL DISEASE, chapter 8.

The diagnosis is established by recovering the virus in the aborted fetus.

BACTERIAL ABORTION

Bacteria that cause abortion ascend into the uterus through the cervix in early pregnancy. Normally the pregnant cervix is impervious to bacteria. This is not the case, however, if a mare has an incompetent cervix. In addition, some mares have secondary ovulations of pregnancy at 40 to 70 days' gestation. During this time, the cervix relaxes and opens. Another group of mares at risk are those with pneumovagina and urovagina who have a chronically infected cervix that appears to be less effective in stopping the passage of bacteria.

Mares who acquire a venereal infection at the time of breeding can develop chronic endometritis and harbor bacteria in the uterus. The bacteria multiply during pregnancy and produce either early embryonic death or *bacterial placentitis*. Abortion takes place when the fetus becomes infected or the placenta is no longer capable of supporting the foal. Occasionally a foal is born alive but succumbs to neonatal sepsis.

Streptococcus zooepidemicus is responsible for most cases of infectious abortion. This bacteria a is a normal inhabitant in the clitoral fossa of mares and appears commonly on the stallion's penis. The infection ascends through the cervix, as revealed by the intense inflammation surrounding the point of attachment of the placenta in this area *(cervical star)*. Abortion takes place between the 5th and 10th months of pregnancy.

Salmonella was once a major cause of epidemic abortion. Infrequent outbreaks caused by *S. typhimurium* still occur. Infection is acquired by contact with contaminated feed and drinking water and infected placental fluids. Abortion occurs in late gestation. A characteristic feature of salmonella abortion is retention of the placenta. The fetus and placenta should be handled with extreme care to prevent transmission of disease to horses and humans. A preventive vaccine is available. It is used infrequently because outbreaks are rare.

Leptospirosis is a disease caused by a spiral-shaped bacteria that attacks both humans and animals. It is spread by contact with the infected urine of rodents, wild animals, and domestic sheep and cattle. Infection in mares, often mild or inapparent, precedes abortion by about 2 weeks. The aborted fetus frequently appears jaundiced (yellow). Bacteria can be shed in the urine of infected mares for 2 to 3 months after recovery. They should be isolated during this time. Antibiotics are not effective in eliminating the carrier state.

Brucellosis is a disease of cattle that causes sporadic abortion in mares. The usual source of infection is contact with soil contaminated by cattle. Pastures used by infected cattle should not be grazed until at least 3 months after the cattle have been removed.

Treatment of the Mare Following Abortion: Cultures should be taken from the endometrium of the mare. These reports will not be available for 2 or 3 days but may be used to modify treatment later. The uterus is infused with an antiseptic or antibiotic. Many veterinarians prefer to infuse the uterus with 2 percent Betadine solution, starting on day 1 and continuing daily for 3 days. Betadine is effective against a broad range of bacterial species, including *Streptococcus zooepidemicus*. After recovery, the endometrium should be recultured to verify that the infection has been eliminated.

The prevention of abortion is like that described for EED.

FUNGAL ABORTION

Aspergillus fumigatus and other fungi account for about 15 percent of placental infections. The mechanism of transmission is like that described above for bacterial *placentitis*. Fungal infections, however, produce less inflammation; as a result, the abortion or stillbirth usually occurs quite close to term. The characteristically appearance of the placenta is thick and leathery.

TWIN PREGNANCY

Although more than one ovulation occurs in approximately 10 percent of estrous cycles, twins occur in only about 1 percent of pregnancies. Twin pregnancy is considered highly undesirable. The mare's uterus simply does not have the surface area to support two placentas. The statistical outcome of twin pregnancies is as follows: Among 100 mares, 30 will abort both foals. Of the remaining 70 mares, 60 will abort one foal and carry the other to term. This leaves 10 mares that will carry both foals to term. However, in 8 out of 10 cases, at least one foal will be born dead, while the surviving foal is often weak, undersized, and may need to be put down for conformation defects. Accordingly, the chance that a mare will deliver two healthy foals is less than 1 percent.

Certain mares are prone to twin pregnancies. Large breeds, including Thoroughbreds and draft horses, have the highest incidence of double ovulation, while small breeds such as ponies have the lowest. Quarter Horses and Arabians are in-between. Mares that have had double ovulations in the past are twice as likely to have them in future cycles.

Mares have a biologic mechanism that operates to prevent twin pregnancy. Recall that embryos wander in the uterus for 16 days before becoming fixed at the base of one of the uterine horns. Ultrasound studies in early twin pregnancies have shown that embryos of different ages and sizes (embryos from ovulations several days apart) tend to fix in the same uterine horn, while those of similar size and age fix with equal frequency in either uterine horn. Thus combining both cases, the majority of embryos fix in the same uterine horn. When two embryos fix together in the same location, the spontaneous elimination of one embryo occurs 70 percent of the time. This is because their membranes, which serve in nutrient exchange, develop a surface-to-surface contact that is nonfunctional and results in the death of one embryo by day 40. The viability of the surviving embryo is not diminished.

In contrast, when two embryos fix at different locations, their membranes are not in contact. Thus they are able to survive beyond the embryo stage to become fetuses. As the two fetuses develop, however, their placentas gradually expand into the mare's uterine body and ultimately make contact. This leads to the gradual weakening or death of one or both fetuses.

Diagnosis of Twin Abortion. Abortion or stillbirth of one or both twins generally occurs at 6 to 9 months' gestation. When both fetuses are aborted, the diagnosis is apparent by observing the two fetuses. The fetuses may be of unequal size, indicating that death occurred at different times. However, when only one fetus is found, the diagnosis can still be made from the appearance of the placenta. The aborted placenta shows a large, smooth area lacking chorionic villi, corresponding to the surface shared with the other placenta.

Treatment of Twin Pregnancy: It is important to diagnose twin pregnancy before 30 days' gestation for reasons explained below. Postbreeding ultrasound exams starting on day 14 postovulation will diagnose most (but not all) twin pregnancies. Rectal palpation is usually not accurate until 2 weeks later.

After diagnosis, a decision must be made whether to immediately terminate the pregnancy or allow it to continue. One reason to allow the pregnancy to continue is to see if the mare will spontaneously reduce the pregnancy to one fetus. If this does not occur by 30 days' gestation, steps should be taken to (a) induce abortion as described below, or (b) convert the pregnancy to a singleton by manually crushing one of the vesicles.

The advantage of inducing abortion is that the mare can be bred on the next heat cycle. If the decision is made to terminate the pregnancy, it should be terminated before the 36th day of gestation. If pregnancy is terminated after that date, the endometrial cups will continue to produce eCG and prevent the mare from returning to estrus for 2 to 3 months (see PSEUDOPREGNANCY, chapter 2).

Selective termination of one embryo is accomplished by transrectal palpation. The embryonic vesicle is grasped through the wall of the rectum and either pinched between the fingers and thumb or forced between the ultrasound probe and the pelvis. Selective termination is relatively easy to do in the first 2 to 4 weeks of gestation, particularly when embryos are located in opposite horns or can be separated from one another before crushing. Nonetheless, there is always a chance that the procedure will result in the loss of both embryos.

Treatment of twin pregnancy beyond 40 days is complex owing to the influence of the endometrial cups, and also because the conceptus is now a fetus. It is difficult to manually crush one fetus while preserving the other. In most cases, both fetuses will be lost. An ultrasound-guided procedure has been used with some success to terminate one fetus in midpregnancy. It involves injecting potassium chloride into the fetal heart. This procedure might be considered if a mare is found to have two live fetuses in midpregnancy.

Several other options are available. The pregnancy can be allowed to continue with a 60 percent chance that the mare will abort one fetus and deliver one live foal. However, both fetuses may die, resulting in a late-term abortion that could interfere with the next breeding season.

The pregnancy can be electively terminated by inducing abortion as described later in this chapter. The mare will return to estrous in 120 days. Although she usually cannot be bred during the same breeding season, she should be ready to breed at the start of the next season.

Prevention: Monitoring twin-prone mares with ultrasonography during estrus will identify those with multiple preovulatory follicles. When two follicles are the same size and look as though they may ovulate at about the same time, one option is to postpone breeding until the next estrus. If it appears that the two follicles will ovulate at different times, it may be possible to avoid twins by waiting 12 to 18 hours after the first follicle ovulates before breeding the mare, at which time the egg from that follicle should no longer be viable. However, note that mares who ovulate more than one follicle are among the most fertile. Strategies that restrict breeding in these mares appear to lower overall pregnancy rates.

The other approach is to breed all mares, regardless of ovulatory function, and manage those with twin pregnancy according to the findings. With this approach, there is no need to screen mares before ovulation. Instead, high-risk mares (or all mares, for that matter) are screened *after* breeding.

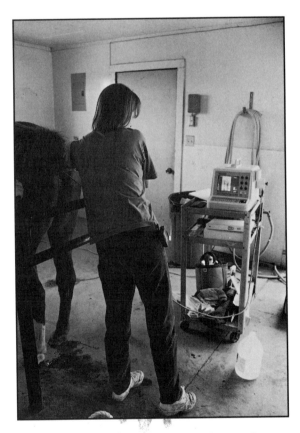

Ultrasound examinations starting on day 14 of postovulation will diagnose most twin pregnancies. This allows treatment to begin before day 30.

The first screening sonogram is done at 14 days' postovulation. At least one additional sonogram is done before day 29. If only one embryo is detected on both exams, the likelihood of twins is remote. If two embryos are detected on the first exam, treatment depends on whether the embryos are adjacent or nonadjacent. If they are fixed in the same uterine horn and cannot be separated, they are observed to see if one will reduce spontaneously. If the embryos are in opposite horns or can be separated easily during rectal palpation, one embryo is crushed at this time.

If the two embryos are still present at the second exam done before 29 days, abortion can be induced with a prostaglandin injection. The mare will return to heat and can be bred on the next heat cycle. Prior to inducing abortion, an attempt can be made to crush one of the embryonic vesicles.

MATERNAL AND PLACENTAL CAUSES OF ABORTION

Premature separation of the placenta, uterine torsion and rupture of the uterus are problems that complicate pregnancy and may cause fetal and maternal death. They are discussed in chapter 13, "Foaling."

INCOMPETENT CERVIX

Cervical incompetence is loss of the protective seal following an old laceration of the cervix. The partially open cervix allows bacteria to ascend into the uterus and infect the placenta. This is a major cause of infectious abortions and still-births.

Mares with an incompetent cervix and a history of midpregnancy abortion may benefit from a procedure to close the cervix. The cervix of these mares is often flimsy and difficult to repair. As an alternative to repair, a pursestring suture can be taken around the cervix to pull it together. This procedure is called a cerclage. The cerclage is placed either 2 days after breeding in anticipation of pregnancy, or after a pregnancy has been confirmed by ultrasonography. The cerclage must be removed before the mare goes into labor.

PLACENTAL DISORDERS

Placental insufficiency is the result of failure of the placenta to develop well formed chorionic villi. The placenta is a mirror image of the endometrial surface to which it attaches. In areas where the endometrium is scarred and has few if any endometrial glands, the corresponding villi are underdeveloped or absent. These areas are evident on inspection of the expelled placenta. Thus placental insufficiency is the consequence of chronic endometritis and endometrial fibrosis (see chapter 8). Placental insufficiency usually causes abortion within the first 90 days of gestation.

Body pregnancy is a rare cause of placental insufficiency. Normally the fetus occupies the body of the uterus and one uterine horn. In body pregnancy, the fetus occupies only the body. The body alone does not have enough interior surface area to support a growing fetus. Abortion occurs when the demands of the fetus exceed the capacity of the placenta. Examination of the expelled placenta reveals an absence of its extensions into the uterine horns.

Hydrops of the fetal membranes is another rare cause of intrauterine death. It is discussed in PREGNANCY COMPLICATIONS, chapter 7.

UMBILICAL CORD ABNORMALITIES

Abnormalities of the umbilical cord have been associated with late-term abortions. A cord may be abnormally long or short. An abnormally long cord lends itself to *twisting*, an event that shuts off the blood supply and causes fetal death from asphyxiation. An abnormally short cord, on the other hand, can tear apart prematurely during labor.

The incidence of fetal death caused by intrauterine twisting of the cord is estimated at about 1 percent. A twisted cord, however, is sometimes blamed for abortion when no other cause is apparent. Thus the actual incidence may be somewhat lower than 1 percent. If death is caused by twisting, the umbilical cord should show swelling and discoloration at the site of the twist.

GENETIC AND DEVELOPMENTAL DEFECTS

Genetic defects are rare causes of abortion. Severe (lethal) defects cause early embryonic death, while less-severe defects generally do not interfere with gestation. Among foals born alive, the overall rate of congenital deformity has been estimated at 4 percent.

The *contracted foal syndrome* encompasses a group of angular limb and/or spinal deformities in which the foal's bones and joints are twisted into abnormal positions. Other malformations, including wry nose and absence of the abdominal wall, may be present. These deformities are believed to be caused by malpositions within the uterus that interfere with normal fetal development. They are not caused by genetic defects. For more information, see FETAL MALFORMATIONS in chapter 14.

TRAUMA AND STRESS

It is rare for kicks and other types of trauma to cause abortion, because the fetus is well-protected by a cushion of shock-absorbing fluid.

Breeding during pregnancy is one exception. Some mares have secondary ovulations of pregnancy at 40 to 70 days' gestation and will stand for breeding. Trauma to the cervix caused by the stallion's penis, along with the simultaneous introduction of bacteria through the relaxed cervix, are the causes of infectious abortion. A similar outcome occurs if the cervix is dilated for the purpose of taking an endometrial biopsy during an unsuspected pregnancy.

Stress is a known factor in mare abortion. Mares who undergo surgery or suffer from colic have a higher incidence of stress-related abortion. Other examples include mares transported under strenuous circumstances; those engaging in intense physical exercise or training; and mares moved from a cool, wet environment to one that is hot and humid. Stress releases steroids and lowers plasma progesterone levels.

TOXIC CAUSES OF ABORTION

FORAGE TOXICITIES

The consumption of certain pasture grasses and poisonous plants has been linked to reproductive problems, including abortion, fetal developmental abnormalities, and prolonged and difficult labor.

Note that toxicity of these plants and grasses is not limited to reproductive disorders.

Among the common forage toxicities are the following:

Fescue Toxicity. Fescue poisoning is caused by a fungus that reproduces in fescue grass during periods of lush growth, usually during the fall after heavy rains. Fescue is less palatable than other pasture grasses. Ordinarily horses will not eat it if something else is available.

Fescue toxicity causes problems during the last 3 months of pregnancy. These problems include delay in the onset of labor, increased frequency of stillbirths, retained placentas, and a reduction or cessation of milk production (see AGALACTIA in chapter 15).

Fescue toxicity can be avoided by not feeding fescue hay or pasture during the last 3 months of pregnancy.

Sorghum Toxicity (Milo, Sudan grass, Johnson grass). Sorghum grasses are found in the Southwest and in much of the eastern United States where they are sometimes used as forages. Pastures are more likely than hay to cause poisoning. Not all varieties are poisonous. Johnson grass is the most toxic; Sudan grass accounts for most cases of poisoning.

Poisonous sorghums contain cyanogenic glycosides that are metabolized to cyanide. Mares consuming sorghum grasses between days 20 and 50 of gestation have aborted or given birth to foals with a variety of skeletal deformities.

To prevent exposure, remove pregnant mares from sorghum pastures during periods of new growth. New growth occurs when a heavy rain follows a frost, after the grass has been trampled, and when a warm spell follows a cool spell.

Ergot Poisoning. Ergot is a fungus that grows on grasses and cereal grains, particularly rye. It appears as a black, banana-shaped mass about one-half inch long that replaces part of the grass or grain seed. Improperly stored grain can become contaminated with this fungus.

In pregnant mares, the continuous consumption of small amounts of ergot over a period of time can cause reproductive problems like those described for fescue poisoning.

Ergot poisoning can be avoided by keeping the seed heads mowed off in the late summer. Grain should be properly stored to prevent the growth of mold and fungus.

Locoweed Poisoning. There are many toxic varieties of locoweed or milk vetch, all of which are members of the legume or pea family. Many horses develop a strong preference for these plants and will seek them out even when better forage is available. Pregnant mares will abort or produce malformed foals if they consume the *astragalus* variety of locoweed in early pregnancy.

Locoism is seen most often on high desert ranges in the western United States during the spring, when locoweed is abundant. Pregnant mares should be kept off locoweed-containing pastures.

Teratogenic Plants. Many plants are known to produce fetal deformities, particularly when consumed by the mare during the first 3 months of pregnancy. Fortunately, most poisonous plants are not very palatable. However, under circumstances such as scarce forage, overcrowding or the introduction of new horses into the herd, pregnant mares may seek out and eat any green plant, including those that are poisonous.

Plants and grasses known to be teratogenic include European or spotted hemlock, lupine, tobacco and hellebore. Those suspected of being teratogenic include autumn crocus, jimsonweed, creeping indigo, wild pea, mimosa, poppies, wild black cherry, groundsel and periwinkle.

To prevent exposure to dangerous pasture plants, dig them out or apply weed killer. Look for these plants along fence rows, ditches and watering troughs. Annual weeds should be mowed before going to seed.

Grass clippings more than a few hours old should not be consumed by horses. The grass ferments and produces sugars that can cause colic and founder. Do not throw clippings and plant cuttings into areas where horses are accustomed to eating hay.

Monitor the quality of the hay being fed and eliminate hay that contains weeds or mold.

DRUGS AND CHEMICALS

Only a few drugs and chemicals are believed to cause abortion in mares. Phenothiazines, certain dewormers, and organophosphates have been implicated in some cases. Because of the possibility of abortion, anthelmintics and insecticides containing organophosphates should not be administered to pregnant mares. In addition, benzimidazole anthelmintics (BZD and Pro-BZD) are not recommended for deworming during the first 4 months of pregnancy.

INDUCING ABORTION

Indications for elective abortion include twin pregnancy, accidental or unwanted pregnancy, and hydrops of the fetal membranes.

Abortion can be induced by prostaglandin injection or intrauterine lavage. Neither method is capable of preventing unwanted pregnancy before the 5th day following ovulation. Uterine infusion is not effective because the embryo is still in the oviduct. Prostaglandin injection is not effective because the corpus luteum is not responsive to *luteolysis* until the 5th day following ovulation.

When possible, elective abortion should be undertaken during the first trimester. After 4 months' gestation, the risk of complications makes it safer to allow the mare to go to term. If abortion must be performed after midpregnancy, intrauterine infusion with large volumes of fluid is the recommended method.

Prostaglandin Injection. A single injection of prostaglandin PGF2α (Equimate™), when given 5 days after ovulation and before the 35th day of gestation, will induce luteolysis and abortion, and return the mare to estrus in 3 to 5 days. This is the easiest method of terminating a twin pregnancy and a pregnancy resulting from a mismating. It is successful in more than 90 percent of cases in early pregnancy.

After 35 days' gestation, Equimate must be repeated daily for 3 to 5 days. Mares who abort after the endometrial cups have been formed will remain in diestrus until the cups cease to function at 110 to 120 days' gestation (see PSEUDOPREG-NANCY, chapter 2). Thus if a mare aborts on day 70, she will not return to estrus and ovulate for about 40 to 50 days.

Intrauterine Infusion. Dilatation of the cervix and infusion of 1 to 2 liters of saline into the uterus is an effective method of inducing abortion and is an alternative to prostaglandin injection after 35 days' gestation. Because the cervix is penetrated, antibiotics are added to the infusion solution to prevent infection. The infusion is repeated if abortion does not occur in 24 hours.

After 80 days' gestation, abortion is usually delayed 2 to 7 days. This can be shortened by rupturing the membranes and gently removing the fetus by traction.

Intrauterine infusion works by irritating the uterus and causing contractions. The irritation also activates the release of prostaglandin from the endometrium. Both mechanisms appear to be involved in causing abortion.

As in the case of prostaglandin, abortion after 37 days is followed by a prolonged period of absent heat.

STALLION

INFERTILITY

Stallion infertility is failure to impregnate a mare. The reasons for such failure include inability to perform the act of breeding for behavioral or physical reasons, and sterility caused by injuries and diseases of the male genital tract.

In stallions, performance causes of infertility are much more common than disease causes. Many performance problems are psychological in origin and can be traced to inexperienced stallion management.

PERFORMANCE PROBLEMS AFFECTING STALLION FERTILITY

LACK OF LIBIDO

Male libido is the driving force behind the stallion's urge to mate. The libido is under the influence of testosterone. It is also affected by the season of the year, as well as psychological stimulation from higher centers in the brain. Not all male horses are born with the same sex drive. Some stallions are intrinsically more vigorous than others. The most powerful and sexually aggressive stallion usually becomes the leader of the herd.

Loss of sexual interest and arousal is the most common behavior problem in stallions. It is a symptom rather than a disease.

Stallions suffering from a low libido generally show little or no interest in estrus mares. They are easily distracted, are slow to get and maintain an erection, and will dismount without achieving *intromission*.

Some factors that may be associated with a low libido are a low serum testosterone level, early-season breeding, sexual overuse, and various forms of sexual inhibition resulting from painful or fearful experiences during mating. These subjects are discussed below.

In autumn and winter, the stallion's testosterone and sperm concentration, as well as his libido, are lower than they will be in April, May and June. Accordingly, the stallion will not be as vigorous or as potent in October through March as he will be at the height of the breeding season.

When a stallion lacks libido, it is important to know what his sexual behavior was like before the problem began. A change in sex drive suggests that something occurred in the interim. Was he injured or frightened during breeding or while displaying stallion behavior? How many mares does he normally service each week? How frequently was he used as a 2- or 3-year-old? Has he ever been used as a teaser? The answers to such questions may point to the reason for the stallion's lack of libido.

A chronic illness associated with fever, debilitation and weight loss can lower testosterone levels as well as sperm production. These parameters begin to improve as the stallion recovers—unless the stallion develops testicular degeneration. Testicular degeneration is the most common medical cause of a low testosterone concentration. It is discussed later in this chapter.

The administration of performance-enhancing drugs can adversely affect the stallion's testosterone concentration and (paradoxically) his libido. More important, these drugs can cause permanent damage to the sperm-producing cells.

Treatment: It is a good idea to obtain testosterone and gonadotropin levels and have your veterinarian perform a breeding soundness examination to rule out medical causes of reduced libido before assuming the problem is psychological.

The sexually inhibited stallion will need to be helped back on track by achieving success in the breeding shed. The best way to accomplish a successful mating is to try the stallion repeatedly using minimal restraint and lots of patient encouragement. Select an experienced mare who is well-liked by the stallion and shows strong signs of estrus, and also who is willing to stand quietly when mounted.

If the stallion does not become aroused, or becomes aroused and loses interest, further efforts at breeding should be discontinued because either horse may become aggressive. Furthermore, lack of success increases the stallion's frustration. Determine whether the mare has ovulated using rectal palpation or ultrasonography. If she remains in heat and has not ovulated, try the horses again in 12 hours.

If success is not achieved with the first mare, consider changing broodmares and breeding locations. Other techniques that have been effective in retraining the stallion include putting him in a stall or paddock where he can watch other stallions during breeding or collection; and allowing him to engage in longer periods of courtship before mounting.

Training the stallion to use an *artificial vagina* (AV) has proven to be a successful method of reestablishing interest in sex. The AV can be made more or less stimulatory by adjusting the temperature and pressure of the inner bladder.

Another method of developing stallion confidence and libido is to turn the stallion out to pasture with one or more experienced broodmares, as described in PASTURE BREEDING, chapter 5.

Finally, even if the stallion's sex drive does not improve, fertilization can still be accomplished through artificial insemination. Neither mounting nor full erection is necessary for ejaculation. A stallion can be collected while standing using manual stimulation and an artificial vagina.

STALLION-HANDLING MISTAKES

Hand breeding requires that the stallion be under restraint throughout the entire breeding process—a situation totally foreign to natural mating. In pasture breeding stallions and mares develop courtship rituals, and are willing and able to breed much more frequently than they are in the breeding shed.

The skilled handler encourages the stallion's self-confidence by allowing him to establish his own courtship and mating routine. Restraint during hand breeding is used wisely and only to enforce safety. Unnecessary discipline, aggressive handling, and the use of punishment to establish dominance are confusing to stallions, disrupt the rhythm of mating, and ultimately produce loss of erection and disinterest in the mare. Over time, these practices foster abnormal sexual behaviors, including lack of libido, fear of mating, and in extreme cases a savage attitude toward mares and even humans.

Stallions in training and competition appear to have a relatively high incidence of sexual dysfunction. It has been suggested that techniques used to discipline stallions on the training circuit are counterproductive when used in the breeding shed. A breeding stallion should be given much greater latitude than would be acceptable in performance training. Because conditions are so different, the stallion may be less confused and may perform better when handled by someone other than his trainer. Many stallions, however, are incapable of handling the duties of breeding and competing at the same time. In these situations, breeding should be postponed until the stallion retires from competition.

TRAUMATIC BREEDING EXPERIENCES

A stallion who has had a painful injury or a severe fright during breeding may develop a fear of breeding. Included in this category are the traumatic experiences associated with rough handling and excessive punishment. Younger stallions are at greater risk of being permanently affected, and also are much more difficult to retrain.

The most common time for a stallion to be kicked by a mare is during the dismount. However, a stallion can be kicked, struck or bitten when the horses make initial contact and at other times during live cover. Understandably, a stallion who has had a painful injury during breeding may hesitate to mount another mare.

Serious injuries to the erect penis (usually caused by a kick) can produce temporary or permanent loss of erection. A poorly padded phantom mare and improper use of the artificial vagina are other causes of penile trauma. Stallion rings and brushes can result in penile abrasions and lacerations. These devices do not serve a useful purpose and should be eschewed (see MASTURBATION in chapter 3). A painful injury that prevents erection can lead to a psychological block that inhibits erection even after the penis is healed.

Traumatic injuries can be prevented by good breeding practices. Mares should be teased across a barrier and not brought to the breeding shed until ready to stand. Injuries during live cover can be prevented by hobbling the mare and handling both mare and stallion as described in COVERING THE MARE, chapter 5.

Treatment is like that described for loss of libido.

OVERUSE

Sexual fatigue is a common cause of loss of libido. A sexually aggressive stallion may be able to service two or three mares a day for several weeks; however, such a schedule would be too heavy for most. In fact, nearly all stallions used extensively at stud show a gradual decline in sexual vigor toward the end of the breeding season. A stale or "sour" attitude can be prevented, in part, by scheduling rest periods. Longer periods are required at the end of the summer as the days grow shorter.

Sexual overuse also can be prevented by booking fewer mares or switching from natural service to artificial insemination (AI). With AI, ejaculatory frequency can be reduced from three times a day to three times a week without affecting the number of mares bred by the stallion.

Note that 2- and 3-year-old stallions used for breeding or semen collection appear to be unusually susceptible to developing abnormal sexual responses, including slow arousal and severe aggression. Unlike older stallions, this behavior does not change after sexual rest. Accordingly it is a good practice not to use very young stallions for breeding; or at least to severely restrict the number of services.

EXCESSIVE TEASING

Stallions used as teasers on a regular basis are susceptible to loss of libido or the development of a negative attitude toward mares characterized by excessive roughness and aggression. Note that continuous teasing without the opportunity to breed results in a decline in sexual arousal in a matter of weeks.

Unless a stallion enjoys teasing mares, it is advisable to use another less valuable stallion for this purpose.

The aggressive behavior of this stallion was caused by using him excessively as a teaser.

EJACULATORY PROBLEMS

Ejaculatory dysfunction is a symptom complex that includes failure to ejaculate, ejaculation outside of the mare, and ejaculation that occurs inside the mare but is incomplete. Incomplete ejaculation is possible because the horse is one of only a few species able to retain a portion of his ejaculate. Incomplete ejaculation causes subfertility because the semen contains a low concentration of sperm.

Stallions with ejaculatory dysfunction exhibit poor or irregular thrusts, thrust repeatedly to the point of exhaustion, dismount without ejaculating, or dismount at the onset of ejaculation. These stallions appear to have a good libido but perhaps associate ejaculation with pain, previous injury or punishment.

Stallions with incomplete ejaculation often show the classic signs of ejaculation, including flagging and urethral pulsations. However, they either do not ejaculate or ejaculate only part of their semen. This problem is difficult to recognize but can be suspected when several mares serviced by the stallion fail to become pregnant. A dismount semen analysis (see COVERING THE MARE, chapter 5) may show sperm. The presence of sperm indicates that ejaculation did occur; however, it does not prove that ejaculation was complete.

Some cases of incomplete ejaculation appear to have a neurological basis and may respond to drugs that contract the smooth muscles surrounding the prostate and accessory sex glands. The treatment of ejaculatory problems related to sexual inhibition is like that described for loss of libido.

MOUNTING PROBLEMS

A stallion with a mounting problem may have difficulty rearing, may be able to mount but not hold the position, or may dismount just prior to ejaculation. Initially this behavior suggests sexual inhibition or ejaculatory dysfunction. However, most cases are caused by pain in the back, hindquarters or feet. Osteochondritis of the hocks, spine or pelvis is a frequent cause. Stallions suffering from chronic laminitis may experience intense hoof-wall pain when standing up on their back legs. Other causes of failure to mount include muscle disease and nerve and spinal cord injuries.

One cause of premature dismount is the presence of sores on the inside of the front legs caused by mounting a phantom mare.

A horse with joint or muscle pain may improve when placed on a *nonsteroidal anti-inflammatory drug* (NSAID) such as phenylbutazone for 1 to 2 weeks.

PREMATURE ERECTION

Premature erection is a condition in which *belling* of the head of the penis occurs prior to intromission, making the penis too large to enter the mare's vagina. The situation can be accommodated by lubricating the vagina with H-R Jelly and manually directing the penis through the vulva.

SEMEN CAUSES OF INFERTILITY

Semen analysis is part of the prebreeding soundness examination of the stallion and is often performed routinely on ejaculates collected for AI. This analysis is also obtained as a first step in the investigation of unexplained infertility. Ejaculates are collected using an artificial vagina.

Ejaculates that contain a low sperm concentration, a high percentage of nonmotile sperm and/or a large number of abnormal-appearing sperm are associated with subfertility. Subfertility may also be due to the presence of blood or urine in the semen.

IMPAIRED SPERM PRODUCTION

Diseases that impair *spermatogenesis* include testicular degeneration, testicular hypoplasia, undescended testicles, acute orchitis, testicular tumor, and testicular injuries followed by glandular *atrophy*. These conditions are discussed below in TESTICULAR DISEASES.

Overuse is a well-recognized cause of a low sperm count. Stallions vary widely in their ability to produce sperm in response to a heavy breeding schedule. Mature stallions produce more sperm with frequent breeding than do young stallions. If a stallion is not exceptionally fertile, excessive use may result in subfertility.

Stallions who received testosterone or anabolic steroids during the course of training or competition are at increased risk of losing their fertility. These drugs act on the pituitary gland to block the release of follicle-stimulating hormone (FSH) and luteinizing hormone (LH). Low plasma levels of these hormones suppress spermatogenesis and also the release of testosterone. Most stallions take several months to recover sperm production after the drug is stopped. In some cases, the condition becomes permanent (see Testicular Degeneration later in this chapter).

A pituitary gland disorder is a rare cause of a low sperm count. If the hypothalamus does not release gonadotropin-releasing hormone (GnRH), or if the pituitary gland is not responsive to GnRH, serum concentrations of FSH and LH will be too low to maintain both sperm production and normal testosterone levels. A low serum testosterone concentration suggests this possibility, but an exact diagnosis requires a series of blood tests available only at schools of veterinary medicine or centers specializing in equine practice. Stallions who have this condition may benefit from injections of GnRH.

Vitamin A is required for spermatogenesis. A severe deficiency results in a low sperm count and/or an excessive number of abnormal-appearing sperm. Vitamin A deficiency is unlikely to occur unless a stallion has consumed a substandard ration for a prolonged period. Such a deficiency is easily corrected and prevented by administering a vitamin A injection and feeding green hay or forage. A stallion whose vitamin A concentration is normal will not benefit from supplemental vitamin A. Furthermore, overdosing with vitamin A has been associated with fragile bones, a rough hair coat, skin disease and internal bleeding.

Vitamin E supplements are occasionally given by horse breeders to enhance reproductive performance. However, there is no evidence that supplementing with vitamin E improves the stallion's libido or the quality of his semen. The same is true for B-complex vitamins.

Stallions found to have impaired sperm production should be sexually rested while awaiting the results of veterinary diagnosis.

HEMOSPERMIA

Blood in the semen is the result of bleeding from the genital tract. It can be recognized by a pink or reddish color in semen collected in an artificial vagina. It is an uncommon cause of stallion infertility.

Hemospermia occurs most often in stallions who have a high frequency of ejaculation. Pain on ejaculation, and signs of ejaculatory dysfunction such as mounting and dismounting, may or may not be present.

A common cause of hemospermia is bacterial urethritis caused by equine venereal disease, discussed in chapter 8. Other specific causes include infected and bleeding cuts and sores of the glans penis and urethral process, strictures of the urethra (most often caused by the pressure of stallion rings), growths of the urethra, stones in the urethra, and infections of the accessory sex glands. The most common cause of a urethral process laceration is a hair from the mare's tail that gets caught between the head of the penis and the vulva. These cuts often become secondarily infected and ulcerate. Cutaneous habronemiasis of the glans penis and urethral process is another cause of ulceration. These conditions are discussed later in this chapter.

Blood in the ejaculate affects the fertility of that ejaculate. The infertility is caused by the presence of red cells rather than serum or some other blood component in the semen. The mechanism by which red cells cause infertility is unknown. The total sperm count, motility, and appearance of the sperm are not affected. Some stallions experience hemospermia intermittently and thus are sometimes fertile. When blood is present in every ejaculate, however, the stallion is either infertile or highly subfertile.

The first step in evaluating hemospermia is to inspect the stallion's penis and urethral opening for inflammation or a bleeding site. The entire urethra, including the openings of the accessory sex glands and the neck of the bladder, is best examined using a flexible fiber-optic endoscope. The procedure is performed with the stallion standing and tranquilized. Positive findings include urethral inflammation, ulcerations, growths, stones, and a bloody discharge from the openings of the accessory sex glands. Ultrasonography of the accessory sex glands and cytological examination of urethral secretions after ejaculation can provide additional information.

Treatment: Ulcers and infections of the urethra often heal following sexual rest for 2 to 4 weeks. Your veterinarian will also prescribe a course of broad-spectrum antibiotics selected on the basis of bacterial culture and sensitivity tests. Drugs that acidify the urine, such as ammonium chloride, have been used successfully, either alone or in combination with antibiotics. Hemospermia caused by urethral strictures, growths, or *lesions* of habronemiasis may require surgery.

Stallions who do not benefit from sexual rest may be cured by an operation called a *urethrostomy*. This operation creates a new opening high in the urethra. Urine passes through this opening, thus bypassing the lower urethra. The surgical opening closes spontaneously in a matter of weeks.

UROSPERMIA

Urospermia is an infrequent cause of infertility. Urine in semen severely depresses sperm motility and results in the death of sperm cells. Nearly all cases are caused by involuntary urination during ejaculation. The reasons for the involuntary urination usually are not known. An occasional case may be associated with disease affecting the spinal cord or the nerves controlling the bladder. The condition is most likely to occur when the bladder is full.

Urine in semen can be suspected by the yellow color and typical odor of the ejaculate. It can be confirmed by dipping a urea or creatinine test strip into the semen. Urospermia is an intermittent phenomenon. In affected stallions, only about 30 percent of ejaculates contain urine. Therefore it is often necessary to inspect several ejaculates in order to make the diagnosis.

Treatment: Drugs that contract the bladder sphincter or prevent bladder emptying during ejaculation have been tried with limited success.

The most effective method of controlling urospermia is to encourage the stallion to empty his bladder just before breeding. One way to accomplish this is to exploit his natural marking behavior by exposing him to a pile of manure from an estrus mare or another stallion. In refractory cases, the bladder can be emptied by inserting a catheter, although this runs the risk of introducing infection.

BACTERIA IN SEMEN

Cultures of the urethra and semen frequently reveal bacteria. The majority of them are not disease-producing. Occasionally a stallion is found to be harboring a bacteria known to be associated with infection—usually *Klebsiella pneumoniae, Pseudomonas aeroginosa, E. coli* or *Beta Hemolytic Streptococcus.*

However, merely harboring bacteria, even a pathogenic species, does not adversely affect the stallion's daily sperm output, number of progressively motile sperm, or his ability to fertilize the mare—as long as the bacteria are not causing an infection. When bacteria are causing infection (such as urethritis and accessory sex gland infection), you would expect to see blood, inflammatory cells, and clumps of purulent material in the ejaculate; a heavy growth of bacteria on the culture plate; and bacteria present on two or more cultures. The motility of the sperm may be diminished, particularly in semen examined 1 hour after ejaculation.

DISEASES OF THE PENIS AND URETHRA

PHIMOSIS, PARAPHIMOSIS, AND PRIAPISM

Phimosis is the inability to protrude the penis from the sheath. *Paraphimosis* is the inability to withdraw the penis back into the prepuce. *Priapism* is a persistent erection without sexual arousal.

Phimosis is usually caused by a sheath infection that results in a stricture that prevents the tip of the penis from protruding. The glans penis inside the infected sheath becomes inflamed and infected. The larvae of Habronema stomach worms, and tumors of the sheath, are other causes of stricture.

In paraphimosis, the penis is dropped and often paralyzed. Penile infection, trauma to the penis, and spinal cord inflammation are common causes of dropped penis. The administration of phenothiazine tranquilizers for surgery and other purposes has been followed by a refractory type of penile paralysis.

In priapism, the penis is not only extended but also tumescent. Causes of priapism include spinal cord inflammation; complications of castration; penile trauma with clotting of the erectile tissue; and constriction of the shaft of the penis, which occurs when the extended penis pushes through a stricture of the prepuce. The constriction prevents blood in the erectile tissue from returning to the body. In spinal cord disease, there is an imbalance between the nerves that control the flow of blood into and out of the erectile tissue (corpus cavernosum), resulting in an unnatural state of tumescence.

Treatment: Phimosis is treated by cleaning the sheath pocket with dilute hydrogen peroxide solution and applying a topical antibiotics, as described in INFLAMMATION OF THE PENIS AND SHEATH on the following page. This may allow the prepuce to open up. If not, it can be opened surgically.

For a horse with a paralyzed penis (paraphimosis), a most important initial step is to manipulate the penis back into the sheath cavity. Ice packs and pressure are used to slowly reduce swelling as the penis is worked back in. Once this has been accomplished, the swelling will start to subside. A pursestring stitch can be taken through the sheath to prevent the penis from dropping back down.

A penis that cannot be retracted into the sheath cavity should be supported using a penile sling and suspensory apparatus tied around the horse's body. If the supporting cup is made of a mesh material such as nylon or polyester webbing, drainage of urine is facilitated. The sling should be removed and washed frequently.

Paraphimosis caused by phenothiazine drugs can often by reversed by a drug called benztropine mesylate. To be effective, the drug must be given within the first 24 hours of detecting paraphimosis.

The treatment of priapism caused by stricture of the prepuce involves surgically dividing the stricture. Priapism caused by clotted blood in the penis can be treated by flushing the corpus cavernosum with heparin solution. The procedure is done under general anesthesia. If the condition persists for more than 4 days, an operation to shunt blood out of the corpus cavernosum may be considered.

When a penis remains extended over a period of time, either because of paraphimosis or priapism, the erectile tissue becomes scarred. Elasticity is lost, and the condition becomes permanent. The skin becomes excoriated and ulcerated. Surgical procedures that may be of benefit include circumcision, a penis-retraction operation, and partial amputation of the penis. The choice depends on the condition of the penis.

PENILE INJURIES

Some specific causes of injury to the penis include kicks to the erect penis by mares during breeding, lacerations of the glans penis or urethral process by a mare's tail hair, improperly fitting stallion rings, and falls on top of jumps and fences.

Following major trauma, there is severe swelling. The penis may become paralyzed and drop down. Bleeding and clotting can occur within the spongy erectile tissue (corpus cavernosum) and result in priapism. If the skin is broken, the penis is quite likely to become infected.

Treatment: With severe trauma, it is important to determine if the urethra is intact. Your veterinarian can determine this by passing a catheter into the bladder.

Clean lacerations of the urethral process are usually repaired at the time of injury. Badly traumatized wounds are left open and allowed to heal around an indwelling catheter. Periodic cleansing and the application of topical antibiotic ointment to open wounds helps to prevent infection. Control of flies is essential.

The immediate postinjury objectives are to control *edema* and prevent infection. Apply ice compresses for 15 minutes three to four times a day for the first 48 hours. An NSAID such as flunixin meglumine (Banamine™) relieves pain and inflammation. Lasix (a diuretic) is often given to mobilize fluid from the site of injury. Swelling due to gravity can be reduced by massage, careful application of an elastic bandage, and hydrotherapy for 20 minutes twice daily.

Hydrotherapy is cold water delivered to the site of injury using a shower head or nozzle. The irrigation also keeps the wound clean and maintains local hygiene. A topical ointment such as A & D™ helps to prevent excessive drying or maceration of the skin. Prophylactic antibiotics may be indicated to prevent secondary bacterial infection. Daily exercise in the form of walking or jogging helps to reduce swelling. Stallions with all types of penile injury should be sexually rested until the injury is healed.

INFLAMMATION OF THE PENIS AND SHEATH

An excessive accumulation of smegma and bacteria within the prepuce and urethral diverticulum can lead to inflammation of the glans penis and surrounding skin. Inflammation also can occur when the resident flora is destroyed by excessive washing of the penis with antiseptics such as Betadine or chlorhexidine.

The skin of an inflamed penis and prepuce appears fluid-filled and swollen. The penis may be dropped.

Treatment: Cleanse the penis and sheath with warm water using a mild soap such as Ivory Liquid™; then apply a topical antibiotic ointment such as Triple Antibiotic Ointment™. If the penis is infected (red, swollen, painful to the touch), contact your veterinarian; a more serious problem may be present.

EQUINE COITAL EXANTHEMA

This mild venereal infection produces blisters, pustules and ulcerations on the prepuce and shaft of the penis. For more information, see EQUINE VENEREAL DISEASE in chapter 8.

HABRONEMIASIS

Larvae of the Habronema stomach worm can be deposited on the sheath or skin of the penis by stable and horseflies. These summer sores (habronema

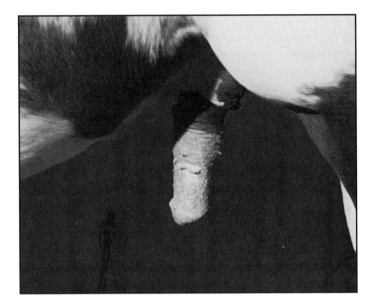

An accumulation of smegma can lead to inflammation of the penis and sheath.

Clean the penis with warm water and mild soap. Rinse thoroughly to remove all traces of soap. Repeat treatment as necessary.

granulomas) often assume a growth-like appearance, resembling a sacroid or squamous cell carcinoma. The lesions often disappear in the winter but reappear and increase in size as warm weather returns.

An open wound or sore during the fly season that suddenly enlarges, ulcerates, and becomes covered with reddish-yellow tissue that bleeds easily should be suspected of being a summer sore. Over the raised round surface of the wound is a gritty, greasy-looking discharge containing rice-sized, yellow, calcified dead larvae. Habronema granulomas often involve the urethral process. They may partially obstruct the urethra and cause spraying and difficult urination.

Treatment: Skin preparations containing topical organophosphates in dimethyl sulfoxide can be massaged into larger open sores with good results. Your veterinarian can provide you with a suitable prescription. Granulomas involving the urethral process require surgical removal.

Summer sores respond well to oral ivermectin paste, which is effective against both the adult worms in the stomach and the larvae in the wound. The inclusion of ivermectin in a deworming program will reduce habronema infections.

PENILE TUMORS

Squamous cell carcinoma is the most common growth affecting the penis and prepuce. Smegma accumulation is a predisposing cause. The tumor begins as a rough, thickened patch of skin or a raised, cauliflower-like growth that ulcerates and becomes secondarily infected. Squamous carcinoma is potentially *malignant*, but it usually does not *metastasize* until it has been present for some time.

An erosive squamous cell carcinoma of the glans penis.

A second common tumor of the penis and prepuce is the *melanoma*—a black-pigmented growth seen primarily in old, gray horses. Melanomas grow slowly, remaining localized for many years. Malignant transformation can occur with rapid spread to vital organs.

The most common *benign* tumor of the skin of the penis is the sarcoid, usually found in horses younger than 4 years of age. Sarcoids have various appearances. Some are rough and wart-like, while others are slightly raised nodules.

Sarcoids have been found by DNA studies to be caused by the cattle wart virus. Although sarcoids do not spread, they have a high rate of local recurrence after surgical removal.

Papillomas and lipomas are other tumors of the penis.

Biopsy is the best way to make the diagnosis. Treatment options include freezing (cryotherapy), a circumcision-like removal of skin, or removal of the entire penis—depending on the size, location and malignancy of the tumor.

URETHRITIS

Sexually transmitted infections are the most common cause of urethritis in the stallion (see EQUINE VENEREAL DISEASE in chapter 8). Sheath infections, habronema granulomas and growths of the penis are other causes.

Signs of urethritis (not always present) are urethral discharge and frequent, painful urination. A stallion with urethritis may pass blood in his ejaculate (see HEMOSPERMIA earlier in this chapter). The diagnosis is made by cultures of the urethra taken before and after ejaculation.

Treatment: The stallion should be sexually rested and placed on appropriate antibiotics for 10 to 14 days. Cultures should be repeated before returning the stallion to breeding. Persistent infection suggests accessory sex gland involvement (discussed below).

URETHRAL OBSTRUCTION

Strictures and stones are the most common causes of urethral obstruction.

Strictures occur after penile injury, improperly fitting stallion rings, urethritis, and tumors of the penis and sheath. A common cause of urethral stricture is a habronema granuloma involving the urethral process.

Smegma can accumulate in the urethral fossa in sufficient amounts to press on the end of the urethra and cause obstruction. This is called a *bean*. It is more common in older horses. The chief sign is spraying.

Bladder and urethral stones are not common in horses. The majority occur in middle-aged and older geldings. The long, narrow urethra of the male prevents the passage of most urethral stones. These stones tend to lodge high in the pelvis just below the outlet of the bladder.

Signs of urethral obstruction include urine spraying; dribbling; dropping the penis; and frequent, difficult urination. Complete urethral obstruction is characterized by severe distress, with groaning and rolling. If a complete obstruction goes unrelieved, rupture of the bladder and peritonitis will ensue.

Treatment: A stricture of the urethra that interferes with urination requires surgical repair.

An obstruction caused by a bean can be treated by manually removing the bean and cleansing the sheath cavity and urethral fossa using mild soap and water. Periodic application of neomycin in vegetable oil prevents recurrence.

Diagnosis of a urethral stone can be made by rectal palpation and/or a transrectal ultrasound. With a complete blockage, your veterinarian will attempt to pass a catheter into the bladder. This relieves the obstruction and may dislodge an impacted stone. However, in most male horses with a complete blockage, a catheter won't pass. The horse will need to undergo an operation to open the urethra and remove the obstructing stone.

Treatment of a ruptured bladder in an adult horse depends on the size of the tear and whether the horse is able to urinate after relief of the blockage. If the horse is stable, antibiotics alone may suffice. Signs of peritonitis indicate the need for immediate surgery.

ACCESSORY SEX GLAND INFECTION

The accessory sex glands are the paired vesicular glands, the paired bulbourethral glands, and the bi-lobed prostate (see illustration *Stallion Reproductive System* in chapter 3). The ampullae are the enlarged terminal 4 inches of the spermatic ducts that carry sperm from the testicles to the urethra. The ampullae empty into the urethra close to the vesicular glands.

The accessory sex glands and the ampullae discharge fluid into the urethra prior to and during ejaculation. Bacteria gain access to the accessory sex glands through the urethra. Thus urethritis and accessory sex gland infection often coexist.

Accessory sex gland infection is important primarily because of its role in causing stallion infertility. The semen is often blood-tinged and contains numerous white blood cells (PMNs), bacteria and clumps of pus. Sperm motility is reduced.

Seminal Vesiculitis. The vesicular glands are thin-walled, tubular structures measuring 6 to 8 inches in length and 2 inches in diameter. They are situated on either side of the bladder neck. They fill with fluid when the stallion becomes sexually excited.

Stallions with seminal vesiculitis usually are asymptomatic, although painful or difficult ejaculation may be associated with some cases (see EJACULATORY PROBLEMS earlier in this chapter). Palpation per rectum occasionally reveals a vesicle that is enlarged and firm.

Ampullary Infection. Infection of the ampulla can be associated with a blockage of the duct at its junction with the urethra. Other predisposing conditions are orchitis, epididymitis and seminal vesiculitis. Rectal palpation may reveal an enlarged ampulla.

The first step in diagnosing accessory sex gland infection is to obtain bacterial cultures from the sheath, penis, urethral fossa, preejaculatory fluid and semen; and the urethra before and after ejaculation. A pure growth of bacteria in the semen, and/or the urethra following ejaculation, is consistent with accessory sex gland infection.

Secretions from the vesicular glands and the ampullae can be examined and cultured separately using a sterile urethral catheter passed to the level of the duct openings. The glands are massaged per rectum and the expressed secretions withdrawn through the catheter. Direct visualization and sampling of the interior of the vesicular glands can also be accomplished using a fiber-optic endoscope.

Treatment: Antibiotics given for 2 to 4 weeks are the treatment of choice for seminal vesiculitis. The stallion should be sexually rested for at least 2 months to prevent relapse. Cultures should be repeated to determine if the infection has been cured before allowing the stallion to return to breeding.

Treatment failures are common. It is difficult to achieve concentrations of antibiotics in vesicular glands high enough to destroy all bacteria. The direct instillation of antibiotics using a flexible endoscope has been described.

Stallions with infection of the ampullae often improve with ampullary massage. The stallion is first stimulated by exposure to an estrus mare. The massage then compresses the distended ampulla and opens a blocked duct. Regular collection of semen may prevent recurrence.

Even when accessory sex gland infection cannot be eliminated, pregnancy can still be achieved by infusing the mare's uterus prior to breeding with an antibiotic-containing semen extender that sterilizes the ejaculate. The antibiotic should be selected on the basis of sensitivity testing. For natural service, 100 to 150 mL of antibiotic-containing extender is infused into the mare's uterus immediately before breeding. For artificial insemination, the extender is added to the semen fraction 30 to 60 minutes before insemination.

Prostate and Bulbourethral Glands. These glands are less commonly involved than are the seminal vesicles and ampullae. Treatment is the same as that described for seminal vesiculitis.

TESTICULAR DISEASES

ORCHITIS

Orchitis is swelling and inflammation of the testicle. It may involve one or both testicles and is often accompanied by *epididymitis*, which is inflammation of the coiled tubules on top of the testicles. Epididymitis can also occur as a consequence of accessory sex gland infection.

Trauma to the testicle from a kick during breeding is the most common cause of orchitis. A severe blow results in hemorrhage and marked swelling of the testicle.

Later the testicle undergoes atrophy and becomes small and hard. Testicular infection can develop if the skin is broken.

Bloodborne bacteria, viruses (EVA, EIA) and the migratory larvae of *strongyles* are infrequent causes of orchitis.

The principal sign of orchitis is a firm, warm, painful swelling of the scrotum. The scrotum may be too swollen to be able to feel the testicle as a separate structure. An ultrasound can be helpful in visualizing the testicle and distinguishing orchitis from torsion of the spermatic cord.

A stallion with acute orchitis usually is febrile and may appear lame. Examination of the ejaculate discloses large numbers of white cells and occasionally bacteria. Sperm motility may be decreased and the percentage of abnormal-appearing sperm cells may be increased.

Testicular degeneration is a common sequel to acute orchitis. Another complication is the development of antisperm antibodies. These antibodies may destroy healthy sperm in the opposite testicle.

Treatment: Early vigorous treatment to control scrotal temperature and reduce swelling is essential in order to prevent testicular degeneration. Apply ice packs every 3 to 4 hours. Support the testicle with a scrotal sling. Hydrotherapy with cold water delivered via a shower head for 15 minutes every 2 to 3 hours is also effective. Anti-inflammatory drugs such as phenylbutazone (Butazolidin™) or flunixin meglumine (Banamine™) relieve pain and suppress inflammation. The stallion should be sexually rested until fully recovered.

Intravenous broad-spectrum antibiotics are given for bacterial orchitis, and may also be indicated to prevent secondary bacterial infection in testicular swelling caused by trauma. If cultures are available, antibiotics are selected on the basis of sensitivity tests. Antibiotics for bacterial infection are continued for at least 1 week beyond the disappearance of pain and swelling.

A severely damaged testicle that becomes atrophic is unlikely to produce sperm and may interfere with the health of the undamaged testicle. Your veterinarian may advise removing that testicle.

TORSION OF THE SPERMATIC CORD

A sudden twist of the spermatic cord of 270 to 360 degrees interferes with the blood supply to the testicle and produces a painful mass in the scrotum similar to that of acute orchitis. The testicle will die unless the problem is corrected within a few hours. Spermatic cord twists tend to occur in young stallions immediately after strenuous exercise.

Palpation of the testicle reveals that the epididymis is out of position, being at the front instead of the back of the scrotum. Ultrasound of the scrotum helps to make the diagnosis.

Treatment: Your veterinarian should be notified immediately on finding a swollen, painful testicle. A twisted cord can sometimes be unwound by twisting the testicle in the opposite direction. When this cannot be accomplished within the first few hours of torsion, surgical correction can be considered if there appears to be a chance of salvaging the testicle. In many cases, irreversible damage will have already occurred and testicular atrophy is the outcome.

A chronic form of torsion exists in which the twist is 180 degrees or less. It may be found incidentally on physical examination. This condition is of no consequence and does not affect the stallion's fertility.

TESTICULAR DEGENERATION

Testicular degeneration is a major cause of infertility in stallions. It is an acquired condition in which the sperm-producing cells of the testicles are damaged or destroyed. The extent of injury varies widely. Some stallions have mild, temporary subfertility; others are permanently sterile.

Testicular degeneration has numerous causes, but elevated body temperature is a common denominator. In order for the testicles to produce sperm, the scrotal temperature must be at least 2 to 3 degrees below that of normal body temperature. Fever caused by a variety of illnesses raises body temperature and therefore scrotal temperature. The problem is further compounded if the horse lies down, which brings his testicles up close to his body. Febrile diseases commonly associated with temporary infertility are equine infectious anemia, pneumonia, laminitis, equine viral arteritis, equine influenza, strangles and ehrlichiosis.

Orchitis and severe testicular injury is often followed by testicular degeneration.

A type of testicular degeneration that can be either reversible or permanent is that caused by the administration of anabolic steroids and testosterone. Early administration during the first year of life, and continuous administration over the stallion's racing career, appear to increase the likelihood of irreversible damage.

The diagnosis of testicular degeneration is based on veterinary examination and semen evaluation. With advanced degenerative changes, the testicles are small and softer than normal, often measuring less than 80 mm (3 inches) in total scrotal width. These findings may suggest the diagnosis of testicular hypoplasia. However, in testicular hypoplasia, the testicles never fully developed; whereas in testicular degeneration, the testicles were normal in size before they began to get smaller. The epididymis of a degenerated testicle may appear to be disproportionately large. This is because the epididymis does not shrink as the testicle shrinks.

The ejaculates of horses with mild to moderate testicular degeneration contain a reduced sperm count and a decrease in the percentage of motile sperm, both in the range of one-third to one-half of normal. The number of abnormal-looking sperm is increased. In severe testicular degeneration, the ejaculate contains few if any sperm cells.

A biopsy of the testicle confirms the diagnosis. This is not always done because of the danger of hemorrhage, degeneration, and development of antisperm antibodies.

Treatment: There is no specific treatment for the degeneration itself. However, prompt treatment of underlying causes may limit degeneration and in subfertility allow the testicles to recover spontaneously. Sexual rest is an absolute necessity. There is no benefit in giving hormones.

Recovery is monitored by periodic semen analysis. It takes at least 60 days for new sperm to reach the ejaculate, so improvement will not be seen for several months. Full recovery can take 4 to 5 months. Feed a ration high in protein and vitamin A, such as fresh legume hay or green pasture.

TESTICULAR HYPOPLASIA

This is a developmental disorder in which one or both testicles fail to reach normal size by maturity. Note that small testicles are the norm in colts and young stallions. The testicles do not achieve adult size until age 4 and continue to grow until age 7. In a normal 3-year-old stallion, the total scrotal width should be greater than 3 inches. If it is less than 3 inches, hypoplasia is likely.

Hypoplastic testicles have a soft, flabby consistency. The ejaculate will show either no sperm or a low concentration with numerous abnormal-appearing sperm cells. The diagnosis can be confirmed by a testicular biopsy.

Testicular hypoplasia should be distinguished from testicular degeneration. The outlook for future fertility is better with testicular degeneration.

INGUINAL AND SCROTAL HERNIAS

An inguinal hernia is the protrusion of intra-abdominal fat or a loop of intestine through the abdominal ring into the inguinal canal. A scrotal hernia is an inguinal hernia that descends through the inguinal canal and enters the scrotum.

Most inguinal hernias occur in colts, either at birth or shortly thereafter. There is a hereditary predisposition for delayed closure of the abdominal ring in most cases. These hernias are considered congenital.

Acquired hernias develop in mature horses as a result of trauma, exertion, and for unknown reasons. In stallions, these hernias can appear quite suddenly following strenuous breeding.

Hernias in the inguinal canal can be detected by feeling a bulge or swelling. In many cases, the hernia can be reduced by pushing it back up through the abdominal ring. If it cannot be pushed back into the abdomen, the hernia is said to be incarcerated. Incarcerated hernias are more serious because they are associated with a high incidence of strangulation. Strangulation occurs when the bowel becomes pinched or twisted, interfering with its blood supply and leading to perforation and peritonitis. Strangulation is a surgical emergency.

A scrotal hernia produces a large mass in the scrotum that resembles acute orchitis or a hydrocele. On rectal palpation, however, the hernia can be recognized because intestine can be felt passing through the abdominal ring.

Treatment: Small inguinal hernias in colts and fillies usually correct spontaneously. It some cases, it is necessary to reduce the hernia on a daily basis to allow the abdominal ring to close. Incarceration is rare.

Surgery is performed on an urgent basis if the foal develops signs of colic or the hernia cannot be reduced. If the hernia is unusually large or does not close within 4 months, elective hernia repair is advised.

Scrotal hernias and inguinal hernias in stallions frequently strangulate and should be repaired as soon as possible. Signs of strangulation in the adult horse are severe and unrelenting colic. Sweating is pronounced. The horse rolls from side to side and often lies on his back. Dehydration and toxicity occur rapidly.

The repair of inguinal and scrotal hernias involves closure of the ring and removal of the testicle on the side of the hernia. If the hernia is congenital, many veterinarians suggest removing the opposite (unaffected) testicle to complete the gelding. In either event, horses with congenital hernias should not be used for breeding because the condition is heritable.

HYDROCELE

A hydrocele is an accumulation of fluid in the scrotum that produces a swelling resembling a scrotal hernia. It is rare in stallions.

The cause of hydrocele in horses is unknown. In some cases, there appears to be a congenital opening in the abdominal ring that allows fluid to trickle from the abdomen into the scrotum. Trauma is another possibility.

The diagnosis can be made by palpation, ultrasonography, and aspiration of fluid from the scrotum using needle and syringe. If the fluid is bloody, a traumatic basis can be suspected.

Treatment, if necessary, involves removing the testicle on the side of the hydrocele.

VARICOCELE

In this condition, the spermatic vein is elongated and coiled back on itself like a varicose vein, forming a nonpainful "bag of worms" in the scrotum. The resulting warm mass raises the scrotal temperature and can interfere with sperm production in both testicles. A varicocele can clot, in which case an inflammatory mass develops that is difficult to distinguish from an acute orchitis.

The diagnosis is made by scrotal palpation. Ultrasonography can be used to visualize the varicocele and rule out other conditions.

Treatment involves removing the varicocele and associated testicle. It is indicated to improve fertility in the remaining testicle, and to treat a clotted varicocele.

UNDESCENDED TESTICLES (CRYPTORCHIDISM)

Testicular descent from the abdomen to the scrotum occurs in utero at about 300 days' gestation. An interruption in the normal sequence of events can leave a testicle in the abdomen (10 percent of the time) or in the inguinal canal (90 percent). A horse with one testicle missing from the scrotum is said to be *unilaterally cryptorchid*. If both testicles are missing, the horse is *bilaterally cryptorchid*.

The rings in the abdominal wall close during the first few days after birth. If a testicle has not passed through the inguinal ring by 2 weeks of age, that testicle will remain permanently in the abdomen. A testicle that makes it through the ring but is retained in the inguinal canal may yet descend. This condition is called *temporary inguinal retention*. It is found predominantly but not exclusively in ponies. If the testicle does not descend from the inguinal canal into the scrotum by 3 years of age, the condition is called *permanent inguinal retention*.

Undescended testicles produce testosterone but not sperm. A horse with two undescended testicles has all the behavior characteristics of a stallion but is infertile. A horse with one undescended and one normal testicle is fertile but should not used for breeding because the condition is heritable.

An undescended testicle in the inguinal canal usually can be felt by palpating the inguinal canal with the horse either standing or anesthetized and lying on his back. A testicle located adjacent to the inguinal ring in the abdomen can often be felt by rectal palpation. Ultrasound is useful in locating testicles in sites where they can't be identified by palpation.

Treatment: A cryptorchid horse should be gelded. During surgery, it is essential to find and remove both testicles. This can be technically difficult if the location of the undescended testicle(s) has not been determined before surgery, and if the testicles are located in the abdomen.

There is a specific syndrome (*proud cut*) in which a horse is said to have been gelded but continues to act like a stallion. The cause is either incomplete gelding or failure to find and remove a cryptorchid testicle. For more information, see PERSISTENT STALLION BEHAVIOR (PROUD CUTS), chapter 3.

TESTICULAR TUMORS

Testicular tumors are uncommon in the horse. The principal sign is gradual enlargement of one testicle, which may go unnoticed for some time. Bilateral tumors are rare.

The most common tumor of the testes is the seminoma. Seminomas are relatively *benign* but may become malignant and *metastasize* to the liver and elsewhere.

The second most common tumor is the teratoma, also called a dermoid cyst. These growths contain hair, cartilage and bone. They occur most often in undescended testicles but also occur in normal testicles. The teratoma is almost never malignant, and in most cases it is discovered as an incidental finding during gelding.

Lipomas are smooth round or oblong growths composed of mature fat cells surrounded by a fibrous capsule. Lipomas occur wherever fat cells exist throughout the body. They grow slowly and may get to be several inches in diameter. Lipomas are only a problem when they can't be distinguished from other tumors.

The diagnosis of a tumor can be suspected on finding an enlarged, painless testicle that is freely moveable within the scrotum. An ultrasound can provide additional information and is useful in ruling out other causes of scrotal swelling. A testicular biopsy using a syringe and small-caliber needle is the best way to obtain tissue for diagnosis.

Treatment: Testicular tumors grow slowly and are usually cured by removing the diseased testicle and its spermatic cord.

The effect of the tumor on sperm production is variable, ranging from a normal semen analysis to infertility. If the quality of the semen is good and the stallion is booked, the surgery is often delayed until after the breeding season.

ASSISTED
REPRODUCTIVE
TECHNOLOGY

EMBRYO TRANSFER

Embryo transfer (ET) involves a set of procedures to remove an embryo from the uterus of a donor mare and place it into the uterus of a recipient mare.

ET may be the only way to obtain a foal from a mare who cannot carry the foal herself. Such mares include those with chronic endometrial fibrosis, reproductive conditions associated with early embryonic loss, and mares with physical problems such as laminitis or healed pelvic fractures that prevent them from carrying or delivering a foal.

ET is an option for obtaining a foal from a mare on the racing or show circuit without having to interrupt her career. It is also a way to obtain a number of offspring each year out of a genetically superior mare. Finally, embryos can be frozen (or placed temporarily into a rabbit host) for shipment to another country.

The overall success of embryo transplant depends on (a) the ability to recover a fertilized egg from the donor mare, (b) the quality of the ovum recovered, and (c) the fate of the embryo after transplantation into the recipient's uterus.

Expected success rates using mares of normal fertility are a 50 to 70 percent embryo collection rate per cycle and a 50 to 70 percent chance of pregnancy after transfer. Thus the overall efficiency of ET per cycle is between 25 and 50 percent.

Currently ET is accepted by the majority of horse registries. However, some registries impose restrictions. For example, in some cases only one foal is eligible for registration per year. If you are considering ET, be sure to check the current policies of your breed registry.

COLLECTING THE EMBRYO

The donor mare is bred or inseminated in the usual manner. Ultrasonography is essential to determine the exact day of ovulation, since this is the date on which all subsequent procedures are based. Typically the embryo is collected on day 7 or 8 after ovulation.

A customized equine uterine flushing catheter is inserted into the uterus through the cervix using a sterile technique. A prepared lavage solution is infused into the uterus by gravity flow. The solution is then allowed to run back out into a filter cup. Fluid in the cup is poured into a search dish and examined under the microscope to find the embryo, which at this stage is about $^1/_2$ mm to slightly more than 1 mm in diameter. If the embryo is not found on day 7, the flush is repeated on day 8.

The quality of the embryo is then graded under the microscope. The embryo is washed several times and placed in a sterile petri dish containing a transfer medium. The embryo is now ready for transfer. This should be accomplished within 3 hours of collection.

Embryo recovery rates are influenced by a number of factors, including the fertility of the mare and stallion. Mares with subnormal fertility, for example, have recovery rates of less than 30 percent.

GRADING THE EMBRYO

The size and developmental stage of the embryo progress rapidly during each 24-hour period. Embryos collected on day 6 are one-third the diameter of embryos collected on day 8. Embryos collected on days 7 and 8 produce the highest pregnancy rates, although they do not withstand freezing as well as embryos collected on day 6. Embryos collected before day 6 and after day 8 have a low incidence of producing pregnancy.

Embryo quality is graded on a classification system that ranges from 1 to 5. Grade 1 embryos are considered excellent, with a uniform round size and intact capsule. Grade 2 embryos have minor imperfections; Grade 3 embryos have definite but not severe problems; Grade 4 embryos are poor with severe problems; and Grade 5 embryos are unfertilized or dead. The grade greatly determines the pregnancy rates. In one series, pregnancy rates were 69 percent for Grades 1 and 2, while those for Grades 3 and higher were 18 percent.

SELECTING THE RECIPIENT MARE

The recipient mare should be between 3 and 10 years of age and approximately the same size as the donor mare. If the recipient is not a maiden mare, she should undergo a complete breeding soundness examination, including palpation and ultrasonography of the reproductive tract, a vaginal speculum examination, and a culture and biopsy of the endometrium. A suitable recipient mare should ovulate or cycle regularly and have no history of infertility.

To ensure that a recipient mare is available and in the right stage of the estrous cycle when the embryo is collected, most ET programs use more than one recipient mare. In addition, it is almost always necessary to synchronize the estrous cycles of the donor and recipient mares in order to ensure that timing is correct (see SYNCHRONIZING OVULATION in chapter 2). For maximum pregnancy rates, recipient mares should ovulate 1 day before to 3 days after the donor mare ovulates.

Recipient mares are examined 5 days after they ovulate by ultrasonography to look for fluid in the uterus and determine the size of the *corpus luteum* (CL). Mares without fluid who have a tightly closed cervix, a tubular uterus and a CL greater than 30 mm in diameter are candidates to receive a transfer.

Ovariectomized mares (mares who have had their ovaries removed) can be used in place of intact mares as embryo recipients. The advantage of using an ovariectomized mare is that the estrous cycle becomes irrelevant. The uterus can be artificially prepared to receive and nourish the embryo simply by administering injectable or oral progesterone for 5 to 7 days before transfer. Thus the need to synchronize ovulation and follow the mare with daily ultrasounds is eliminated. Pregnancy rates for ovariectomized mares are similar to those for intact mares.

TRANSFERRING THE EMBRYO

There are two methods for transferring the embryo. One is to make an incision in the mare's flank under IV sedation and local anesthesia. The uterine horn then is brought through the incision and a small incision is made in the horn close to the ovary. The embryo is drawn up into a glass pipette and transferred into the uterine horn.

The second method involves inserting the tip of a pipette into the body of the uterus through the cervix. The transcervical approach was found to have lower pregnancy rates, perhaps because of the introduction of bacteria and contaminants into the uterus during the procedure. With improvements in technique and more experience on the part of technicians doing transcervical transfers, pregnancy rates with transcervical transfer are considered by many to be equal to those of implantation into a uterine horn.

COOLED-TRANSPORTED AND FROZEN EMBRYOS

The development of culture media that preserve embryos for up to 48 hours makes it possible to move embryos within and between countries. Special handling of the embryo and preparation of the culture media is essential. The cooled embryo can be stored and transported in the same cooling unit used to transport cooled semen (e.g., the Equitainer system).

Pregnancy rates with cooled-transported embryos transferred into recipient mares within 36 hours are similar to those obtained by the transfer of fresh embryos.

Freezing of embryos is not widely employed. Embryos older than 6 days have poor viability after freezing and thawing. The advantage of freezing an embryo is that the genetic potential of the mare can be preserved indefinitely. Pregnancy rates are significantly lower with freeze-thawed embryos but may improve with further research.

IN VITRO FERTILIZATION

In vitro fertilization (IVF) has been accomplished successfully in horses but is still considered experimental.

In the IVF procedure, an egg from a donor mare is placed in a petri dish with sperm from a stallion. After fertilization, the resultant embryo is grown in culture and transferred into a recipient mare as described above.

Applications of IVF could be extended by using fertility drugs to produce multiple ovulatory follicles. Each follicle could be aspirated using a needle and syringe. Thus several eggs could be harvested during a single estrus cycle. Although some of this has been accomplished, further applications depend on the progress of ongoing research.

PREGNANCY PREVENTION

When a stallion is kept on the same premises with mares, unwanted breedings can occur. The risk of accidental pregnancy can be lessened by keeping the stallion in a box stall or in some other enclosure with secure fencing. A stallion outside of his enclosure should be under the control of its handlers at all times.

An intact male who will not be used for breeding is often gelded. This is not automatic. Some people prefer to ride and work a stallion. Gelding, however, does eliminate the risk of accidental breeding with that individual.

The removal of both ovaries (ovariectomy) is a major operation and is rarely done for the purpose of preventing pregnancy. A hysterectomy (removal of the uterus) is not done for sterilization in horses.

There is just one "birth control" preparation currently available that will prevent a mare from coming into heat (see PREVENTION OF ESTRUS below).

If accidental or unwanted pregnancy does occur, the available options are to induce abortion (see INDUCING ABORTION, chapter 9) or allow the mare to carry the pregnancy to term.

REMOVING THE OVARIES (OVARIECTOMY)

The principal reason for removing the ovaries is to eliminate estrus behavior in mares not intended for breeding. Ovariectomy also may modify the aggressive behavior of mares with stallionlike behavior (see chapter 8). Some nymphomaniac mares may improve after an ovariectomy. However, by the time the ovaries are removed, the behavior may have become fixed.

Occasionally an ovariectomized mare is used instead of an intact mare to collect semen for artificial insemination. The mare is often given estrogen to ensure signs of heat.

Mares who have had both ovaries removed are excellent embryo transfer recipients because it is not necessary to manipulate their estrous cycles.

The presence of an ovarian tumor is the principal reason for removing a single ovary.

The ovaries can be removed by a vaginal, flank or midline abdominal incision, depending on the size of the ovaries, reasons for surgery and preferred method of the veterinarian.

The vaginal approach is the most convenient, because it requires a minimum of operative equipment and can be done under IV sedation and local anesthesia with the mare standing in stocks. It also has the fastest recovery rate. It is limited by the size of the ovaries and the experience of the surgeon.

The flank approach can also be done under IV sedation and local anesthesia with the mare standing in stocks. The exposure is limited for removal of the opposite ovary. The flank approach may leave an unsightly scar.

The midline approach is the best choice for extremely large ovaries. The operation is done under anesthesia with the mare lying on her back. The surgery may be followed by complications associated with prolonged recumbency.

To prepare the mare for an ovariectomy, withhold feed (but not water) for 36 to 48 hours prior to surgery. Some veterinarians give preoperative mineral oil by stomach tube. This clears out the colon and creates more room for the surgery.

After surgery, antibiotics and anti-inflammatory drugs such as phenylbutazone and flunixin meglumine are prescribed to prevent infection and relieve pain. Recovery usually is smooth. Most mares are able to return to full activity in 3 weeks.

Complications following surgery are unusual. They include intra-abdominal bleeding and shock during the first 24 hours, wound infection, rupture of abdominal contents through the incision, and peritonitis.

PREVENTION OF ESTRUS

Altrenogest (Regu-Mate™) is a progesterone drug that mimics the effects of a persistent corpus luteum. While taking the drug, the mare will not come into heat or exhibit estrus behavior. Eliminating heat makes it easier to handle some racing and show mares. In addition, some mares perform better when not in heat.

The dose of Regu-Mate required to suppress estrus is 0.044 mg/kg body weight per day. In a 450 kg mare, the daily dose would be 9 mL of the 0.22 percent solution in oil. The drug can be administered directly into the mouth using a dose syringe or added to the daily grain ration. Estrus behavior disappears 2 to 3 days after treatment is started and does not return until treatment is stopped.

Progesterone weakens the cellular defense mechanisms that protect the endometrium and may predispose the mare to uterine infection. Accordingly, it is common practice to stop the drug from time to time and allow the mare to cycle. It is important to consult with your veterinarian before starting a mare on altrenogest.

The liquid Regu-Mate is readily absorbed through the skin and can cause serious problems in people. All individuals should avoid direct contact with this drug. This is particularly true for pregnant women, women of child-bearing age who may be unaware of a pregnancy, women who have had complications with oral contraceptives, and men and women with a history of clotting disorders (thrombophlebitis, stroke, coronary occlusion). As with prostaglandins, if accidental contact does occur, the skin should be thoroughly washed with soap and water.

FOALING

GETTING READY

To establish a time frame for completing the prefoaling arrangements, the breeding date can be used to roughly estimate the mare's expected date of delivery. However, even with accurate breeding dates, it is not possible to precisely pinpoint the day of foaling because of the extreme variation in the length of normal gestation in the mare (a range of 327 to 357 days).

If the mare is going to foal in the pasture, there should be enough space so that she can withdraw from the herd as foaling approaches. She will want to do this by instinct. If adequate space is not available, other horses may harass or even injure the newborn foal during the time when the mother is unable to protect it. The mother and foal will return to the herd when the baby is up and about.

CONFINEMENT

Most breeders prefer to confine the mare where she can be closely watched during late pregnancy and *parturition*. The mare should be introduced to her place of foaling at least 4 weeks prior to delivery. The foaling quarters should be clean, dry, draft-free and warm—and preferably in familiar surroundings away from strange people and other distractions. A spacious box stall is the best confinement, but a small, dry, grassy paddock can be equally satisfactory. A good light source should be available since foaling generally occurs at night.

The floor of the stall should be covered with several inches of bedding. Clean straw provides good bedding and firm footing. Sawdust, wood shavings and sand are poor bedding materials associated with the potential for postpartum infections. Change the bedding at least once a day and maintain a clean, sanitary floor.

During the day, the mare should be placed in a large paddock or taken out of her stall and exercised twice daily. At night, she should be confined to her stall.

A mare who has had a Caslick's operation for pneumovagina should have the vulva surgically opened to prevent tearing of the perineum during delivery. This should be done 1 month before her expected due date.

The materials you will need in preparation for foaling are the following:

- A bucket of warm water and a mild liquid soap, such as Ivory™
- Cloth strips for tying up the umbilical cord
- Bandage scissors
- Cotton string for tying off the umbilical stump in the event of bleeding
- A bulb syringe to clear the foal's nostrils if breathing is delayed
- Sterile, disposable surgical gloves
- Several tubes of K-Y Jelly
- Paper towels for washing the mare's perineum
- A flashlight with fresh batteries
- 2 percent iodine (3 ounces) to saturate the umbilical stump
- A cotton blanket or bed sheet

As foaling time approaches, it is essential to check on the mare more frequently. During the night, she should be looked at with a flashlight every 15 to 20 minutes. Stall-mounted, closed-circuit television monitors have been used successfully on many breeding farms.

SIGNS OF APPROACHING LABOR

At 2 to 3 weeks before foaling, the mare's abdomen becomes noticeably enlarged and drops. There may also be swelling beneath the skin along the midline of the abdomen in front of the udder. This condition, called ventral edema, is due to fluid accumulation in the soft tissues. It tends to occur most often in older, multiparous mares.

During the last 2 weeks of gestation, the sacroiliac ligaments relax and the muscles located on each side of the tail head soften and sink, causing a hollowed-out depression in the region of the croup. Also during this period, the mare's vulva relaxes and elongates.

Softening and relaxation of the cervix (noted on vaginal speculum exam) indicates impending parturition. However, the cervix may remain firmly closed until the end of the first stage of labor or may relax and dilate several weeks before. When the mucus plug is expelled, the mare may exhibit a small amount of vaginal bleeding. In most cases this "bloody show" is not observed.

The enlarged and dropped abdomen of the pregnant mare near foaling.

Changes in the Udder. At 2 to 4 weeks before foaling, the mare's udder begins to enlarge, swelling at night and shrinking during the day. This distension is most noticeable in the *multiparous* mare. The *teats* or nipples of the breast become noticeably enlarged 1 week before foaling. During the 24 hours immediately prior to foaling, the udder remains full and tense.

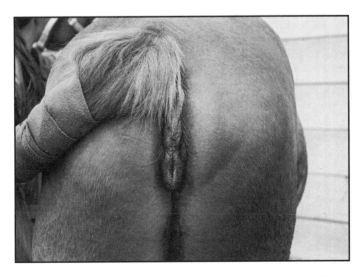

The mare's vulva relaxes and lengthens within 2 weeks of foaling.

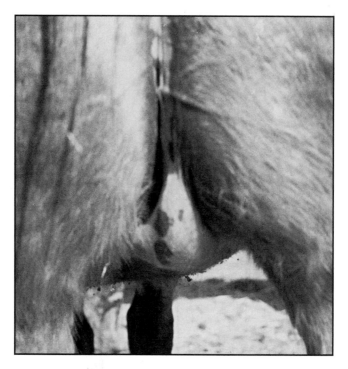

The mare begins to "bag up" 2 to 4 weeks before foaling.

The most important indicator of impending foaling is the appearance and composition of the colostrum produced in the udder. To recognize and follow changes in fluid composition, the mare will need to be milked. If you have not done this before, ask your veterinarian to demonstrate the technique so that you can learn to do it yourself.

Here's how the milking procedure works:

1. Begin with a light massage of the mare's udder to accustom her to the feel of your hand.

2. As the mare relaxes, use the thumb and index finger to gently compress (but not occlude) a nipple close to its junction with the udder. A slight upward nudge will cause milk to flow into the small teat cistern above the nipple.

3. Now pinch off the top of the nipple to keep the milk from returning to the breast.

4. With the remaining two fingers and thumb, encircle the nipple and then squeeze the fingers from top to bottom to express milk into the palm of your other hand.

Sampling the colostrum as described in the text is the most accurate way to predict when foaling will occur.

The quality of the milk changes as labor approaches. Initially the udder secretions are clear and watery. Later they become yellow and slightly more viscous. Still later the secretions become thick, syrupy and amber-colored. In the final stage the milk is thick, white, sticky and opaque. This usually occurs 12 to 24 hours before foaling but may occur as late as 24 hours after the foal is born.

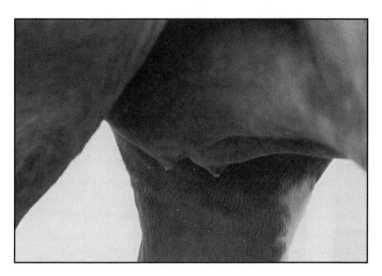

Waxing. A honey-colored bead forms at the end of each teat within 48 hours prior to foaling.

Waxing is the appearance of dried, honey-colored beads that form at the end of each teat. These beads are the result of increased production of sebum from sebaceous glands surrounding the tip of the nipple, along with the passage of small amounts of colostrum from the breast ducts. When waxing is noted, it is quite likely the mare will foal within the next 48 hours.

Note, however, that some mares exhibit waxing and may drip colostrum for up to 2 weeks before foaling. If a mare is losing colostrum before the foal is born, she should be milked out and the colostrum frozen for later use.

Commercially available foaling-predictor kits can be used as a general indicator of when to expect labor. These kits use test strips to detect calcium and other electrolytes in mare's milk. The concentrations of calcium and magnesium increase many fold 48 to 72 hours prior to foaling. The increase in the calcium concentration is what accounts for the change in the color of milk from amber to white.

As the mare's udder enlarges, the skin crease between the two halves of the udder—as well as the folds between the udder and thighs—may become damp and irritated. To prevent skin infection, gently wash the mare's udder with warm water and Liquid Ivory™ soap. Rinse thoroughly to remove all traces of soap. Apply Vaseline™ to protect the skin. Repeat as necessary.

Mares have a degree of control over the timing of labor and prefer to foal at night; 80 percent do so between 10 p.m. and 4 a.m. On the day of foaling, mares usually become restless and spend more time pacing and getting up and down.

Once you have determined that foaling is imminent, wrap the mare's tail with a clean bandage, starting at the base of the tail and continuing to the end of the tailbone. The wrap minimizes contamination of the vulvar area by the mare's tail hair. Wash the vulva and perineum with a mild soap such as Liquid Ivory™ and rinse thoroughly with a hand-held sprayer or paper towels soaked in warm water.

LABOR AND DELIVERY

Although labor is a continuous process, for descriptive purposes it is divided into three stages. The first stage is preparatory. It is marked by contractions that open the cervix and position the foal for delivery. The second stage is the actual birth. The third stage is the passage of the placenta.

Normal delivery may be preceded by several bouts of *false labor* characterized by mild to severe colic, frequent urination, listlessness, kicking at the abdomen and rolling. These signs are like those of the first stage of labor but differ in that they do not progress and the mare returns to normal activity.

Of all births, 95 percent will occur naturally without problems.

STAGE 1

The preparatory stage lasts 2 to 4 hours. Uterine contractions gradually dilate the cervix and shift the foal from the resting position into the orientation for

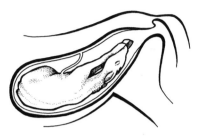

(A) The foal in the resting position before labor begins.

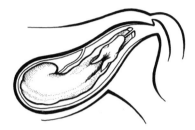

(B) In stage 1 labor, the foal rotates with his head and legs extended into the birth canal.

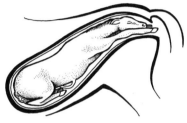

(C) The nose and forelegs pass through the birth canal with the soles of the feet pointing down.

delivery. Even though the mare does not strain, she becomes noticeably more active and restless, paces back and forth in her stall, gets up and down frequently, passes small amounts of urine and stool, and kicks and nips at her abdomen. Patchy sweating is often noted behind the elbows.

Most mares in early labor exhibit some or all of these signs, but occasionally a mare passes through stage 1 without giving any indication that foaling is imminent. This type of mare presents a dilemma in that foaling may occur before her attendants are ready.

STAGE 2

The second stage of labor is a period of active abdominal straining. It begins when the water breaks and ends with delivery of the foal.

Note that the equine placenta is a complicated structure composed of a sac within a sac. The outer sac is called the allantochorion. On the outer surface of the allantochorionic membrane are the microvilli that intermesh with the glandular lining of the uterus. This outer membrane is richly supplied with blood vessels and has a dark red, velvet-like appearance. Inside the allantochorionic sac is a large volume of yellow-to-chocolate–colored allantoic fluid. Also inside the allantochorionic sac is a second sac that surrounds the foal. This sac is called the

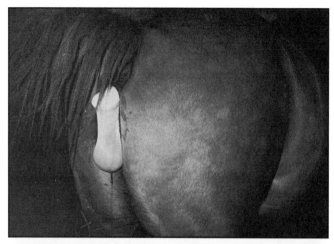

Stage 2 labor. The amniotic bubble protrudes through the vulva.

A normal delivery. The forelegs and head have passed through the birth canal.

The amniotic membrane covers the foal's head.

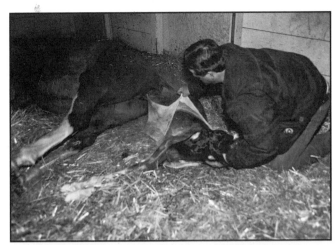

The amnion is removed to facilitate breathing.

The foal rests with his hind legs in the mare's vagina.

The mare stands, breaking the umbilical cord.

amnion. The amnion contains clear fluid. It is not the amnion that ruptures first. Instead it is the outer, allantochorionic sac containing the yellow-to-chocolate–colored fluid that ruptures at the beginning of active labor. Passage of this fluid corresponds to "breaking of the water."

Stage 2 labor typically is brief, lasting 10 to 20 minutes. The mare strains vigorously with powerful contractions of her abdominal wall muscles. These contractions occur in groups of three or four, followed by periods of rest lasting up to 3 minutes.

After the water breaks, the amnion advances to the vaginal opening and presents as a thin, bluish-white, translucent bubble protruding through the vulva. The foal's feet and forelegs are visible inside the sac. The amniotic bubble containing the forelegs should appear at the vaginal opening within 15 minutes after the water breaks. If the amnion ruptures within the birth canal, only the forelegs covered by the translucent membrane will be seen.

In the unusual situation in which the allantoic sac does not rupture at the beginning of stage 2 labor, the mare will attempt to push it out ahead of the foal through a dry birth canal. This results in premature detachment of the placenta from the wall of the uterus. Instead of the bluish-white amnion presenting at the vaginal opening, you will see the red, vascular, velvet-like allantochorionic membrane. This is an obstetrical emergency. The red membrane must be torn open to allow for the release of the allantoic fluid; see PREMATURE SEPARATION OF THE PLACENTA (PLACENTA PREVIA) in chapter 14. After the delivery, be sure to call your veterinarian to examine the mare and the foal for birth-related complications.

Most foals are born within 15 minutes after the water breaks. There is a maximum time between rupture of the amnion and delivery of the foal (usually given as 45 minutes) after which delayed labor is likely to result in fetal death. The reason for this is that the mare's placenta begins to separate in localized areas shortly after she starts to strain. In time, large areas of placental surface are no longer in contact with the endometrium and the foal is not receiving oxygen. Accordingly it is important to record the time when the water breaks, because from that moment forward, the delivery is on a critical time schedule. If the foal is not born within 20 minutes after the water breaks, notify your veterinarian at once!

After the water breaks, the mare usually lies down on her side with her legs extended. Some mares get up and down several times. Rarely a mare does not lie down as her foal appears at the vulva. This is considered undesirable. Not only does the foal drop several feet onto the floor, but also the umbilical cord ruptures prematurely, causing excessive blood loss. Mares who refuse to lie down tend to be nervous, insecure, and easily disturbed by noises, commotion and people. A nervous mare will often lie down if the attendants withdraw.

The normal fetal position in the birth canal is the *anterior longitudinal presentation*. In this presentation, the foal is aligned lengthwise with his spine parallel to his mother's spine. His head and nose lie between extended forelegs with the soles of his feet pointing down (as if he were diving). The tip of his nose is about at the level of his knees.

As the foal's chest enters the birth canal, one foot precedes the other by about 3 to 6 inches. This is very important, because the foal is widest between his shoulders. Extending one leg ahead of the other narrows this diameter. If the legs are even, both shoulders enter the birth canal at the same time. This can result in a SHOULDER LOCK (see chapter 14).

In most cases, the amnion will have ruptured by the time the feet are protruding through the vulva. In the unusual circumstance in which the nostrils are still covered by the amniotic sac, step in quickly and tear or cut open the sac. Otherwise do not interfere with the mare and foal. The mare is perfectly able to provide for her foal and needs no assistance. In fact, excessive and unnecessary interference may lead to postpartum problems that would otherwise not occur.

After the shoulders are delivered, the hips follow easily. Stage 2 labor is now complete. The foal usually rests with his hind legs in the mare's vagina for the next 10 to 20 minutes. During this time the foal receives additional blood from the placenta. Do not sever the cord. This occurs naturally with the movements of the mare and/or the foal.

The umbilical cord normally separates by shredding. Shredding allows the vessels to contract. This is an important mechanism in preventing serious bleeding from the stump. You can expect the navel stump to ooze for up to 1 minute after rupture of the cord. This is not a concern. However, in the unlikely event that a steady stream of blood persists for more than 60 seconds, you should be prepared to tie off the umbilical stump with a piece of cotton string. This is not routinely done, because tying the stump predisposes the foal to cord infection. If it becomes necessary to tie the cord, the tie should be removed as soon as possible (within hours).

Soaking the umbilical stump in 2 percent iodine solution is an important step in preventing neonatal sepsis.

After the cord separates, it is very important to thoroughly soak the umbilical stump with a solution of 2 percent iodine. Repeat this twice during the next 2 to 3 hours. Disinfecting the umbilical stump greatly diminishes the occurrence of naval infection and the potential for neonatal septicemia.

The long end of the ruptured cord remains attached to the amnionic membrane and placenta, which may be seen partially extrude from the mare's vulva. To prevent the mare from stepping on the cord or kicking at the placenta and possibly injuring the foal, the cord and amnion should be tied up well above the level of the hocks with cloth strips.

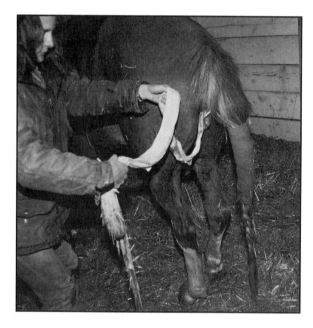

The placenta usually passes within 1 hour. Tie up the cord and amnion well above the level of the hocks.

In the unlikely event that the cord does not break within 30 minutes or the placenta is delivered before the cord ruptures, you should be prepared to step in and break the cord by hand. The cord normally breaks at a predetermined weak point recognized by a slight constriction about 1 to 2 inches from the foal's abdomen. Grasp the cord on the foal's side of the weak point (close to the belly wall) and maintain counter-traction to prevent pulling on the umbilical ring, which could cause an umbilical hernia. With the other hand, grasp the cord on the mare's side of the weak point and pull until the cord separates. Do not cut the cord with scissors or knife. This almost always leads to bleeding and will require tying off the umbilical stump.

The foal should be observed immediately after birth for signs of weakness or depression. The healthy, vigorous foal rolls up onto his chest within 15 minutes and stands within 1 hour. Once on his feet, the foal searches for the udder and begins to nurse. A vigorous foal is nursing within 1 to 3 hours of birth (2 hours is the average). A foal who is not standing by 1 hour or is not nursing by 3 hours is weak and below par. This foal should be seen immediately by a veterinarian and treated as indicated.

For more information on the newborn foal, see chapter 15, "Care of the Newborn Foal."

The foal should be nursing within 3 hours of birth. If not, call your veterinarian.

STAGE 3

The passage of the afterbirth is accompanied by slight-to-mild pain caused by renewed uterine contractions. In most cases, the placenta is expelled within 1 hour of foaling. If the mare does not pass the afterbirth within 3 hours, complications may ensue and treatment is indicated. Notify you veterinarian (see RETAINED PLACENTA later in this chapter).

Do not attempt to deliver a partially expelled placenta by pulling on the umbilical cord. A partially attached placenta cannot be freed by traction. Either the cord will break or the uterus will be pulled inside out.

During separation of the placenta from the lining of the uterus, the apex of the placental sac becomes inverted. As the placenta detaches and rolls down the uterine horn, the placenta turns inside out like a sock. The result is that the velvet-like allantochorionic membrane containing the microvilli that was once on the outside of the placenta is now on the *inside*, while the smooth, rubber-like surface that once faced the foal is now on the *outside*.

Gravid Horn

Uterine Body

Nongravid Horn

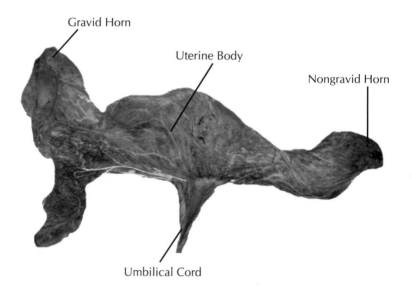

Umbilical Cord

Examine the afterbirth to be sure all surfaces are intact. Arrange the placenta in the Y-shaped configuration it occupied in utero.

The afterbirth should always be carefully examined to be sure that no parts are missing. It is important to handle all fetal membranes with rubber gloves. Inspection is best carried out by arranging the placenta in the configuration it once occupied in the mare. An intact placenta is Y-shaped—a configuration representing the two horns and body of the uterus. Both surfaces should be examined for imperfections. This may require turning the placenta inside out. The most common location for a piece to be missing is at the apex of one of the uterine horns. Surface areas where the rough, velvet-like membrane is smooth represent sites where microvilli failed to separate along with the placenta. Small tears can be detected by filling the placental sac with water and ballooning it out. If there is any suspicion of missing parts, the mare should be examined by a veterinarian.

RESUSCITATING THE NEWBORN FOAL

Immediately after birth a foal should have a breathing rate of at least 60 breaths per minute and a heart rate of at least 60 beats per minute. After 1 hour, the breathing rate should be 30 to 50 breaths per minute, and the heart rate 80 to 130 beats per minute. A newborn foal who does not breathe spontaneously, or gasps and breathes at an irregular rate, is suffering from respiratory depression or an obstructed airway. This foal should be resuscitated without delay.

The first step is to clear the airway of amnionic fluid and possibly *meconium*. The best way to accomplish this is by suctioning the nasal passages. However, in many facilities, a suction machine is not available. The next-best method of clearing secretions is to raise the foal's chest and hindquarters and allow the secretions to run out his nose. Use a bulb syringe to suction secretions at the nasal openings with a bulb syringe, or wipe out the nostrils with a cotton cloth. If you are physically able to do so, hold the foal upside down by his back legs and swing him around in a circle.

If the foal makes no effort to breathe, rub him vigorously all over with a towel and flex and extend his legs. You can also attempt to activate coughing and breathing by tickling the inside of his nostrils with a surgical clamp or a piece of straw. If successful, the foal will gasp a few times and begin to breathe at a regular rate.

If the foal still does not respond, the situation is that of neonatal *hypoxia*. This is a serious problem and must be treated aggressively using assisted breathing and possibly chest compressions. If the foal is flaccid and shows no signs of life, take a moment to quickly determine if there is a heartbeat. If you do not have a stethoscope, place a palm on each side of the foal's chest beneath the elbows and feel for the heartbeat. If no heartbeat is felt, you can attempt to listen for the heartbeat by drawing the left leg foreword and pressing your ear against the side of the foal's left chest behind his elbow. While one person prepares to resuscitate the foal, an assistant should summon the veterinarian.

When a foal is born without signs of life and has no heartbeat, hypoxia has resulted in cardiac standstill. It is unlikely that this foal can be revived. However, when a foal is depressed but there is still some indication of life, assisting the foal's breathing during this critical period can be lifesaving.

The most effective way to administer respiratory support to a newborn foal is through an anesthesia bag connected to an endotracheal tube passed through the mouth or nose into the lungs. Mouth-to-nose breathing is only minimally effective, principally because it does not provide uniform lung expansion. Nevertheless, if endotracheal intubation is not available, performing mouth-to-nose breathing while waiting for the veterinarian may provide marginal support for a limited time to a foal who is not breathing at all. For a foal who is breathing at a very slow rate, mouth-to-nose breathing may be of significant benefit.

The technique for *assisted breathing* follows:

1. Close one nostril with the flat of your hand to prevent air from exiting that nostril.

2. Enclose the other nostril with your mouth. An assistant should hold the foal's lips closed to prevent air from escaping through his mouth. (If an assistant is not available, you will have to do this yourself.)

3. Blow in with enough force to expand the foal's chest.

4. Release both nostrils to allow air to escape.

5. The expiratory phase should be at least twice as long as the inspiratory phase. Continue at a rate of 25 breaths per minute, while pausing every 60 seconds to see if the foal is starting to breathe.

If the foal does not have a heartbeat, or the rate is slow and irregular (less than 20 beats per minute), begin external cardiac massage. This requires at least two people. One continues to ventilate while the other provides chest compressions.

The technique for *cardiac massage* follows:

1. Place the foal with his right side down on a firm surface, preferably with a sandbag under the sternal end of his rib cage.

2. The attendant should kneel beside the foal, placing the heel of one hand over the heart—which is located beneath the fourth and fifth ribs just behind the elbow. Place the heel of the second hand on top of the first. Keep both elbows straight, and push down firmly on the rib cage.

   It requires about 75 pounds of pressure to depress the chest wall a required 2 inches.

3. Quickly release.

4. Repeat at the rate of 60 to 80 chest compressions per minute, allowing one breath for every three compressions.

   Note that chest compression is associated with potential complications including broken ribs and *pneumothorax*.

5. Continue CPR until the foal develops spontaneous breathing and a steady pulse. If this does not occur within 10 minutes, the likelihood of success is remote. Consider stopping CPR.

CARE OF THE POSTPARTUM MARE

POSTPARTUM CHECK

After the delivery, the mare and foal should have a veterinary examination. If all went well, you do not need to call your veterinarian until the morning after the delivery.

VAGINAL DISCHARGE

A reddish-brown vaginal discharge (lochia) is present for the first 3 days and ceases by the 6th day. A discharge that persists beyond 7 days, or one that changes color and/or becomes foul-smelling, is abnormal. The mare should be examined for delayed uterine involution, retained placenta or postpartum metritis.

Note that a mare can have a postpartum uterine infection without an abnormal vaginal discharge. Therefore it is a good practice to monitor the mare's rectal temperature throughout this period. A temperature above 101°F indicates fever and possibly infection.

POSTPARTUM COLIC

During the first 48 hours after foaling, mares often experience episodes of colic caused by cramps associated with uterine involution. In most cases the pain is mild and intermittent, but may be severe enough to cause the mare to thrash and roll.

Note that it is difficult to distinguish normal postpartum colic from the more serious colic that occurs with invagination of the uterine horn, rupture of the uterus and intra-abdominal bleeding. Accordingly, all mares with moderate to severe postpartum colic should be examined by a veterinarian.

HEALTHCARE

All mares should be dewormed 12 hours after foaling with a broad-spectrum anthelmintic that is effective against threadworms, ascarids, and large and small strongyles. Ivermectin is a good choice. Deworming reduces milk-borne transmission of larvae and environmental contamination by parasites and their eggs. Worming thereafter can proceed according to one of the schedules described in chapter 7, "Pregnancy."

Postpartum mares are susceptible to constipation. It is advisable to feed a light-to-moderate diet for the first few days after delivery. Laxative feeds such as bran mash help to reduce the frequency of impaction.

If you plan to breed on the foal heat, it is a good practice to have your veterinarian examine the mare by rectal palpation 1 week after delivery to determine her readiness for breeding.

FEEDING FOR LACTATION

The mare's nutrient requirements during lactation are higher than at any other time in her life. A lactating mare will produce three times her own weight in milk during a 5-month period of lactation.

Nutrient requirements during the first 3 months of lactation are shown in Table I, appendix. In essence, the lactating mare requires almost twice as much dietary energy and protein, and more than twice as much calcium, phosphorus and vitamin A as the mare at maintenance.

When grass hay or forage is being fed free-choice, a 500 kg mare probably cannot consume enough feed to meet these requirements without sacrificing nutritional reserves. However, if you are feeding unrestricted amounts of high-quality alfalfa or lush green legumes, she should be able to consume enough roughage to meet her energy needs. However, she may need additional phosphorus, which can be supplied free choice by a mineral mix such as monodicalcium phosphate. The nutritional requirements for lactation can also be met by feeding 14 pounds of alfalfa and $9^1/2$ pounds of grain per day.

The lactating mare needs 30,000 IU of vitamin A per day. Alfalfa and legume hay (if stored for less than 6 months) may meet these requirements. Corn is the only grain that contains significant amounts of vitamin A. Therefore, in many cases, a supplement is advisable. Vitamin A supplements (and supplemented feeds) are of greatest value when used during lactation.

If you are using a commercial ration as the sole source of feed, note that standard horse rations do not contain enough protein for the lactating mare and often are deficient in lysine—the most important amino acid for high-quality milk production and growth of the foal. If you plan to continue using such a ration, be sure to add a protein supplement. Soybean meal is the supplement of choice, because it is an excellent source of both protein and lysine.

Alternatively, you might want to use a commercial grain supplement for lactating mares that is specifically formulated to be fed along with hay or equivalent forage. These rations meet protein and energy requirements when fed according to the directions provided. Examples are grains designed to be used as creep feeds; or mare and foal pellets.

A salt-mineral supplement such as trace-mineralized salt should be available for free-choice consumption. A nursing mother requires a great deal of water and should have continuous access to a fresh supply.

POSTPARTUM COMPLICATIONS

Foaling complications include a retained placenta, postpartum bleeding, lacerations of the birth canal, invagination of the uterine horn, acute postpartum metritis, rupture of the uterus and prolapse of the uterus. All these complications are much more common after difficult and prolonged labor. A troublesome management problem is the mare who refuses to accept and care for her foal.

Problems with nursing are discussed in chapter 15.

POSTPARTUM BLEEDING

Intra-abdominal bleeding is a serious, often fatal event caused by the rupture of a large artery in the mare's pelvis or abdomen. Rupture usually occurs during foaling but may occur in late pregnancy. The reasons for arterial rupture are unknown. Mares older than 12 years of age are most susceptible. Signs are severe colic, followed by weakness, staggering, pale mucous membranes, shock and collapse.

Surgery in this circumstance has an extremely high mortality rate and is rarely successful. The best hope for survival involves confining the mare to a quiet stall and keeping her as calm as possible. Intravenous fluids and blood transfusions may be given to replace losses. Sedation is often of benefit. The bleeding may stop spontaneously.

Bleeding from the uterus and birth canal can be recognized by the passage of clots and fresh blood from the vagina. This should be distinguished from the normal postpartum vaginal discharge consisting of reddish-brown, old blood.

Manually tearing the placenta from the wall of the uterus is the most common cause of major postpartum hemorrhage. Lacerations involving the cervix, vagina and vulva generally produce a milder form of bleeding that stops spontaneously.

A well-recognized cause of vulvar bleeding is the rupture of a large varicose vein found beneath the vaginal mucosa, often in the area of the hymen. Usually this also stops spontaneously.

Treatment: A mare with postpartum bleeding should receive immediate veterinary attention. Bleeding from the uterus is treated with oxytocin. The treatment of vaginal bleeding caused by lacerations is described below. Intravenous fluids and blood transfusions may be required to treat shock and replace blood.

Cervical lacerations should be repaired 3 to 4 weeks postpartum when the uterus has returned to normal size.

VAGINAL AND VULVAR LACERATIONS

Birth-related injuries to the vaginal *vestibule*, vulva and perineum are most likely to occur when the delivery is complicated by an abnormal presentation.

Most injuries are caused by the feet or muzzle of the foal. The mildest form of trauma is a first-degree laceration that involves only the mucosa of the vagina or the skin of the vulva and perineum. Second-degree lacerations are deeper and involve underlying muscle. In third-degree lacerations, the feet actually tear through the shelf between the vagina and rectum and produce either a rectovaginal fistula or a common opening between the rectum and vagina.

Treatment: If there is doubt about whether the mare was given a tetanus toxoid booster 3 to 6 weeks before foaling, a booster is given at about this time. Tetanus antitoxin also is given if the immune status of the mare is unknown. Prophylactic antibiotics may be indicated.

First-degree lacerations require no treatment. Bleeding is not a problem and healing occurs in a matter of days. Persistent bleeding from deeper lacerations can be controlled with sutures or tampon-like vaginal packs. Most third-degree lacerations should not be repaired at the time of injury because of contamination, swelling, and the presence of revitalized tissue. It is better to cleanse the wound and wait 3 to 6 weeks, or until after the foal has weaned.

The treatment of rectovaginal fistulas is discussed in chapter 8.

RETAINED PLACENTA

The placenta normally passes within 2 hours of foaling. A placenta that is not expelled within 3 hours is considered to be retained. Occasionally the placenta passes but one or more pieces are missing. This usually involves the tips of the placenta attached to the uterine horns. Finally, microvilli may remain attached to the endometrium as the placenta separates from the uterus. This can be recognized by finding smooth areas on the red, velvet-like allantochorionic surface of the placenta. Retained microvilli and placental fragments are important causes of delayed uterine involution.

A retained placenta is visible as tissue protruding from the vulva. It may be only slightly protruding, or it may be hanging below the hocks. In most cases, only a portion of the placenta remains attached to the lining of the uterus. A placenta hanging below the hocks should be tied up to prevent the mare from stepping on the membranes.

A placenta is more likely to be retained after prolonged labor. An increased incidence of retained placenta also occurs in mares over 15 years of age, mares with a history of a retained placenta in the past, C-section mares, and mares who abort after the 7th month of gestation.

A retained placenta delays uterine involution. The major concern, however, is that a retained placenta may lead to an acute postpartum metritis and/or laminitis. Although many mares can tolerate a retained placenta for several days without signs of toxicity, the longer the placenta remains, the greater the likelihood of complications.

Treatment: Notify your veterinarian if the placenta does not pass within 3 hours after the birth of the foal. Most cases of retained placenta respond promptly to the administration of oxytocin. Oxytocin stimulates forceful uterine contractions. The drug can be given intramuscularly or intravenously. Severe uterine cramping may ensue. The dosage, route and frequency of administration will be determined by your veterinarian.

If oxytocin treatment is not successful, the placenta usually can be removed by uterine infusion. In this procedure, the placenta and/or the uterus is infused with several liters of warm Betadine™ solution. The pressure of the fluid causes the uterus to contract and expel the placenta within 30 minutes. Antibiotics are sometimes added to the infusion solution.

Manual extraction involves one of several techniques in which the placenta is physically separated or pulled free from the wall of the uterus. This procedure is associated with an increased risk of major bleeding and the development of acute metritis. Accordingly, manual extraction is recommended only when other treatments have failed.

Prophylactic antibiotics are given to mares with a retained placenta of 24 hours or longer. Other indications for giving antibiotics include a prolonged delivery and a birth canal contaminated by obstetrical manipulation. Nonsteroidal anti-inflammatory drugs such as phenylbutazone an flunixin meglumine (Banamine™)

are given to prevent the circulatory effects associated with the development of laminitis.

Tetanus prophylaxis is important. If the mare did not receive a tetanus toxoid booster 3 to 6 weeks before foaling, and a tetanus toxoid booster should be given now. If her immune status is unknown, she should also be given tetanus antitoxin.

If the mare develops signs of uterine infection, treat as described for acute postpartum metritis.

ACUTE POSTPARTUM METRITIS

This is an uncommon but serious infection of the entire uterine wall characterized by the rapid onset of toxemia, *septicemia* and laminitis. It tends to occur among mares who have had a prolonged, complicated labor; a retained placenta; or extensive bacterial contamination of the uterus.

Signs appear 12 to 36 hours after foaling. They include fever, increased heart and respiratory rates, marked apathy and depression. The vaginal discharge is copious, reddish-brown and foul-smelling. Bacteria and toxins are absorbed into the circulation, contributing to signs of toxemia.

Laminitis can occur at any time and can be severe enough to cause sloughing of the hooves. Signs of laminitis are lameness, warm hooves, pounding digital pulses, and the typical founder stance with the hind limbs placed well foreword under the body to reduce weight on the front legs.

Treatment: It is directed at evacuating the uterus, which usually contains several gallons of infected blood and pus. This is best accomplished by inserting a sterile large-bore stomach tube into the uterus and flushing the cavity repeatedly with copious volumes of warm, sterile water until the recovered fluid is clear. Antibiotics are instilled when the uterus is empty. This procedure is repeated daily for several days. The mare is also given intravenous oxytocin to stimulate uterine contractions, and systemic antibiotics to control infection. Flunixin meglumine (Banamine™) is beneficial for its anti-endotoxic properties, and phenylbutazone (Butazolidin™) for its anti-inflammatory effects. Withdraw grain from the diet and feed a low-carbohydrate, low-protein diet such as high-quality grass hay. Be sure to keep water available at all times.

Treatment of laminitis should be under veterinary guidance. Shortening the course of toxemia with early treatment of the infected uterus reduces the likelihood of laminitis.

DELAYED UTERINE INVOLUTION

Shortly after foaling, the uterus starts to shrink and return to its normal size and shape. This process is called involution. It occurs rapidly. By 1 week postpartum, the uterus has returned to about twice its normal size. Some mares are able to conceive and maintain a pregnancy by the 15th postpartum day. In others, the process takes longer (see BREEDING ON THE FOAL HEAT in chapter 4).

Uterine involution is helped by nursing and exercise. Nursing releases oxytocin, which stimulates uterine contractions. Exercise promotes uterine tone and strengthens the abdominal wall muscles.

A delay in uterine involution suggests the possibility of retained placental tissue as the cause of the delay.

Treatment: The size and shape of the uterus can be determined by rectal palpation. The fluid content of the uterine cavity can be determined by transrectal ultrasound. The presence of fluid indicates that involution is not complete.

If your veterinarian finds that involution is not proceeding according to schedule, he or she may want to administer oxytocin to stimulate uterine contractions. If retained placental tissue is suspected, a uterine biopsy will establish whether this is the case.

RUPTURE OF THE UTERUS

Rupture of the uterus occurs with difficult and prolonged labor, torsion of the uterus, and forced extraction to remove a malformed or abnormally positioned fetus. Occasionally the uterus ruptures for unknown reasons.

The process begins with a tear of the uterine wall, usually caused by the feet of the foal. The laceration may remain small or extend to involve the length of the uterus.

Signs may not be apparent until several hours after foaling. They include severe and continuous colic, often accompanied by vaginal bleeding. If the tear is large, loops of bowel can drop through the uterus into the birth canal and protrude through the vulva. In the absence of this finding, the diagnosis is made by rectal palpation or abdominocentesis. In the latter procedure, a large-bore needle is inserted into the abdomen. The finding of blood and bacteria in the abdomen suggests rupture of the uterus.

Treatment: The mare is treated for shock and dehydration. The tear in the uterus must be repaired. This is best accomplished under general anesthesia through a midline abdominal incision. Peritonitis is treated by washing out the abdomen with large volumes of sterile saline and antibiotics.

If the mare is too sick to tolerate surgery, substitute treatment involves administering ergonovine maleate to contract the uterus. The abdomen is flushed repeatedly using a large-bore catheter until the return is clear. Antibiotics are administered systemically.

Nonsurgical treatment is not as successful as surgery in preventing fatal peritonitis. It is more likely to be successful when the tear is small.

Preterm rupture of the uterus is discussed in chapter 7.

INVAGINATION OF THE UTERINE HORN

In this condition, a horn of the uterus turns inside out and projects into the body of the uterus where it can be felt by rectal palpation. Invagination is sometimes associated with a retained placenta. The invagination occurs as a result of traction on the horn during efforts to expel the placenta. Signs of invagination are colic and restlessness not relieved by pain medication.

Treatment: The placenta (if present) must be manually removed before the horn can be replaced. The horn is then kneaded inward until it returns to its normal position. The infusion of 1 or 2 gallons of warm, sterile water facilitates the replacement.

PROLAPSED UTERUS

This is an uncommon complication in which the uterus turns inside out and protrudes through the vulva. Prolonged straining during and after a difficult delivery, and pulling on the umbilical cord in a misguided attempt to deliver a retained placenta are two common causes.

The uterus must be replaced as soon as possible to prevent shock and other complications. This procedure is quite difficult and requires veterinary management.

Treatment: While waiting for the veterinarian, wrap the prolapse in a clean towel or sheet moistened with warm water to protect the everted uterus from further contamination and drying. Keep the mare on her feet and walking. This slows down contractions and prevents the mare from damaging the uterus by backing up against the wall.

Before your veterinarian can begin to replace the uterus, the mare must be sedated to prevent straining. This is accomplished by sedative drugs and epidural or general anesthesia. The uterus is carefully and thoroughly cleansed with disinfectant soap and then worked back through the pelvic opening until it is completely reduced. The infusion of 1 or 2 gallons of warm, sterile water completes the repositioning of the uterine horns.

Antibiotics are placed into the uterine cavity, and then the vulva is sutured in the same manner as a Caslick's procedure. Intravenous oxytocin is given to shrink the uterus. Antibiotics are administered to prevent postpartum metritis. A prolapsed uterus predisposes to tetanus. If the mare was not given a tetanus toxoid booster during the last 3 to 6 weeks of pregnancy, administer a booster now. Tetanus antitoxin is also given if the immune status of the mare is unknown.

A mare who sustains a uterine prolapse is apt to do so again on the next foaling.

ABNORMAL MATERNAL BEHAVIOR

A mare learns to recognize and care for her offspring as her foal is born, cleansed, and begins to nurse. Hormonal changes during and after delivery sensitize the mare's central nervous system to the sight and sound, and especially the smell and taste of her newborn foal.

For various reasons, a mare's brain may not respond to the usual sensory stimuli. For example, if a mare has had a prolonged labor or a C-section, exhaustion combined with the artificial hospital environment could derail the normal imprinting progress.

Lack of maternal bonding is more common in *primiparous* mares. Some first-time mothers seem neglectful of their foals. They do not lick and nurture them after delivery and frequently wander about in a state of oblivion while the foal makes futile and unassisted attempts to find the teat.

A more troublesome problem is the mare who actively resists suckling and may injure the foal while trying to escape. A distended, painful udder may be the cause of the mare's reluctance to allow the foal to nurse.

Lack of bonding appears to be genetically and/or environmentally determined. This appears to be true for Arabians, who have a higher incidence of foal rejection than other breeds. Studies suggest that many neglectful mothers were rejected foals themselves—or were raised in isolation away from the company of horses.

Occasionally a mare develops an overprotective attitude and sees other animals and people as dire threats. These mares can be dangerous to their attendants. Furthermore, in their intense desire to protect their foals, they may accidentally injure them. Overprotective behavior subsides as the mare gains confidence. The mare and foal should be isolated from other horses for 2 to 3 weeks and disturbed as little as possible until the mare becomes more relaxed.

Rarely a mare attacks and savages her newborn foal, usually biting it on the withers. The aggression occurs while the foal is attempting to stand. The behavior is similar to that of the foal-savaging stallion; accordingly, it has been suggested that it may be caused by masculinization of the mare's central nervous system by unknown mechanisms. Hormone therapy has been attempted with little success.

Treatment: The mare who is oblivious to her foal or refuses to allow it to nurse should be restrained while one or two people help the foal to locate the teat. Once the mare realizes that suckling relieves udder distension, she usually becomes less fearful and readily accepts the foal. This may take several nursing sessions.

If bonding does not occur after suckling, and the mare continues to refuse the foal, she can be restrained in a stall divided by a horizontal bar at shoulder height. The bar is placed so that it holds the mare against the wall but enables the foal to go underneath to nurse. Tranquilizing the mare is not a good alternative. Tranquilizers delay foal acceptance. In addition, they are passed in the milk and can sedate the foal, which results in ineffective nursing.

The aggressive mare who attacks her foal is a difficult problem. Attempts to reunite the foal and dam are rarely successful. The mare and foal must be watched at all times. It is best to remove the foal and raise it as an orphan.

INDUCING LABOR

Inducing labor is not a routine procedure. It carries certain risks, including premature separation of the placenta, malpresentations, retained placenta and the possibility of delivering a premature foal. However, in high-risk mares, the option to induce labor has the overriding advantage of permitting a veterinarian to be present at parturition in the event of life-threatening complications.

High-risk mares are those with a history of premature placental separation, a history of delayed or complicated delivery, mares carrying foals with neonatal isoerythrolysis, mares carrying twins, and those with urgent medical complications. A partial list of urgent medical problems would include preterm colic, edema of late pregnancy, hydrops amnion, premature lactation, and impending rupture of the prepubic tendon. High-risk mares may benefit from transfer to an equine center equipped to handle postpartum complications and to provide neonatal intensive care.

DETERMINING FETAL MATURITY

It is possible to use oxytocin to physically induce labor after 300 days' gestation without regard to whether the foal can survive outside the womb. Unlike human babies, who can be born 2 or 3 weeks early with an excellent chance of survival, the window between immaturity and maturity in the foal is narrow. In fact, foals induced a few weeks early (between 300 and 320 days' gestation) usually do not survive. Even after 330 days' gestation, some mares who are forced to deliver will produce immature foals whose survival is questionable. Apparently some foals are ready to be born sooner than others, and the length of gestation is only one factor in determining the capacity of a specific foal to survive outside the womb.

Accordingly, it is extremely important that delivery not occur before a foal is fully mature. In general, a mare should not be induced until she is close to (or beyond) her expected delivery date—usually not before 330 days' gestation. However, due dates are not always good indicators of fetal maturity because of the variation in length of normal gestation. Signs that labor is imminent (discussed earlier in this chapter) are helpful in confirming an accurate date of delivery. These signs include a dropped abdomen, relaxation of the sacroiliac ligaments, softening and relaxation of the cervix, and changes in the udder and milk secretions—in particular, the appearance of white colostrum.

Commercial foaling-predictor kits (discussed earlier in SIGNS OF APPROACHING LABOR) provide good evidence of fetal maturity and can be used to confirm the mare's readiness to foal. These kits measure changing electrolyte concentrations in colostrum. For example, a high and rising concentration of calcium and potassium, along with a low and dropping concentration of sodium, is indicative of a mature fetus. If the findings are reversed, the fetus is not yet mature and induction should be postponed.

INDUCTION

Oxytocin is the safest and most reliable drug currently in use for inducing labor. This naturally occurring hormone is synthesized in the hypothalamus and stored in the posterior pituitary gland, where it is released spontaneously at the beginning of labor and results in forceful uterine contractions. Oxytocin also initiates the ejection (letdown) of milk from the mammary glands.

Oxytocin is administered intravenously or intramuscularly, according to one of several protocols. Foaling begins shortly after the drug is started and is usually complete within 1 hour.

C h a p t e r 14

DIFFICULT
FOALING

DYSTOCIA

The prolongation of any stage of labor is called dystocia. Most dystocias are fetal in origin. The size or position of the foal creates a blockage in the birth canal that can't be overcome by intense and forceful straining. In about one-third of cases, the blockage is caused by a severely deformed foal.

Certain breeds have a higher rate of occurrence of dystocia. Large draft breeds have the highest incidence (10 percent). Thoroughbreds and Standardbreds have the lowest incidence at about 4 percent.

Unlike most other species, the mare has a rudimentary placenta, classified as diffuse in type. It is attached by thousands of microvilli to the endometrium of the body of the uterus and both uterine horns. During the second stage of labor, small pockets of microvilli separate from the uterus, ultimately resulting in large areas of detachment as the placenta strips away. The foal is dependent on the placenta to supply oxygen. If sufficient placenta detaches before the foal is born, fetal *hypoxia* and death will ensue. In general, if a foal is not delivered within 20 minutes after the water breaks, fetal hypoxia becomes a major concern.

Difficult and prolonged labor is associated with frequent mare complications. They include lacerations of the birth canal, acute metritis, retained placenta, delayed uterine involution, rupture of the uterus and uterine prolapse.

Abnormal presentations are a common cause of difficult or arrested labor. In these situations the cross-sectional diameter of the foal is too wide to pass through the birth canal.

Presentation refers to the alignment of the long axis of the foal with the long axis of the mare—and also to the part of the foal that is proceeding into the birth canal. In a *longitudinal presentation*, the foal's backbone is parallel to the spine of the mother. In a *transverse presentation*, the two spines are at right angles. In an *anterior presentation*, the head and forelimbs are extending into the birth canal. In a *posterior presentation*, the hind limbs or buttocks are the leading parts.

As illustrated in chapter 13, 99 percent of mares deliver foals in the anterior longitudinal presentation. The posterior presentation accounts for less than 1 percent. The transverse presentation is rare (0.1 percent).

Position is the relationship of the foal's back to the four quadrants of the mare's pelvic ring (the sacrum, right and left iliac bones, and pubic bone). In the normal *dorsosacral position*, the foal slides out of the birth canal on his stomach, with his back curving beneath the mare's sacrum. In the *dorsoilial position*, the foal is lying on his side, with his back resting against the right or left iliac bone. In the *dorsopubic position*, the foal comes out upside down, with his back sliding over the pubic bone.

Posture is the relationship of the foal's extremities to his body. In the normal delivery, the foal's extended head and front feet enter the birth canal, with the hind feet trailing behind (as if the foal were diving).

Thus the terminology for the normal foal delivery is *anterior longitudinal presentation; dorsosacral position; with the head, neck and forelimbs extended.*

WHEN TO CALL YOUR VETERINARIAN

Before the mare is due to deliver, be sure to contact your veterinarian and discuss who will be available in the event of an emergency. In mares, difficult deliveries are associated with fetal distress. Any delay in the process of foaling is by definition an emergency! Do not hesitate to call your veterinarian at once.

WHAT TO DO WHILE WAITING FOR YOUR VETERINARIAN

If you anticipate getting help within a few minutes, it is best to get the mare back on her feet and walking. Walking usually delays labor for 30 to 60 minutes.

If your veterinarian will not be available within a few minutes, you may still be able to get help by maintaining phone contact so that he or she can talk you through the problem. This has worked successfully on many occasions.

EQUINE OBSTETRICS

OBSTETRIC TECHNIQUES AND PROCEDURES

The following obstetrical techniques and procedures are used to treat dystocias in mares.

CALL YOUR VETERINARIAN IF...

- The mare is in active labor (straining) for more than 4 hours without rupture of the water bag.
- The water breaks, but the two front feet do not appear at the vulva within 20 minutes.
- The foal is not delivered within 20 minutes.
- You see anything other than the normal presentation, with the head between the forelimbs and the soles of the feet pointing down.
- Progress stops and you see a thick, red membrane at the vulva, indicating premature separation of the placenta. (Act on this immediately.)
- The placenta is not delivered within 3 hours after foaling.
- The placenta appears unhealthy or contains missing pieces.

Repulsion. This is the gentlest and most effective method of assisting a difficult delivery. The objective is to relieve the blockage by repositioning the foal in the birth canal so that the mare can deliver the foal under her own power.

The first step is to push the foal back up into the uterus to gain room to grasp and straighten out the limbs, head and neck. Repulsion must be timed to occur *between* contractions. It is virtually impossible to push a foal back into the pelvis when the uterus is contracting and trying to push it out. Furthermore, attempts to do so may rupture the uterus.

Traction. This involves pulling on the foal's legs to help the mare complete a difficult vaginal delivery. The object is to assist, but not replace, the natural expulsive efforts of the mare. *There is absolutely no indication to pull on a foal when delivery is proceeding in a normal fashion.*

Grasping and pulling on a leg or some other part without first repulsing the foal and relieving the blockage is futile. It also is time-consuming, wedges the foal more tightly into the narrow birth canal, and runs the risk of permanently damaging or killing the mare and/or the foal.

Traction should be applied *only when the mare is straining* (exhibiting abdominal contractions). If the mare stops straining, it will be impossible to pull out the foal. Traction on the legs should be applied in a downward direction toward the mare's hocks rather than straight out beneath her tail. This is to conform to the natural arc of the birth passage.

Moderate to forceful traction is what one or two people of average to above average strength can generate while grasping and pulling on one or more limbs.

Forced Extraction. In forced extraction, obstetrical snares or chains are placed on the limbs or other parts of the fetus with the intent of pulling the foal's head, neck, body or limbs into a position or alignment that lends itself to vaginal delivery. The foal is then extracted by pulling on the chains while the mare is straining.

The decision to use forced extraction should be based on training and experience. Forced extraction can cause irreparable damage to the mare's reproductive tract or result in death of the mare and/or her foal. Inexperienced personnel should not attempt forced extraction because of the likelihood of causing tears and ruptures that they would be unlikely to detect and unable to treat.

Lubrication. Traction becomes less effective with the passage of time. The uterus contracts down tightly around the foal. The pelvic canal, exposed to air and trauma, loses its outer layer of mucous and dries out quickly. Copious amounts of lubricant must be instilled into the uterus to correct this problem.

Commercial water-soluble obstetrical lubricant is available in 2.5 gallon containers. The entire contents of the container can be instilled into the uterus using a sterile stomach tube and a hand pump. Veterinarians carry this equipment with them, and large breeding farms often stock them.

Cesarean Section. Although most cases of mare dystocia can be corrected by obstetrical procedures, a cesarean section (discussed below) may be necessary. C-section should be considered when there is little or no progress after 15 minutes of obstetrical manipulation and the foal is still alive with a chance for survival. The operation, however, must be performed at a veterinary facility. Unfortunately, there may not be enough time to transport the mare to such a facility in time to save her foal.

Fetotomy. If the foal is dead and cannot be delivered by other means, your veterinarian may elect to remove it by fetotomy. With the mare under epidural anesthesia, the fetus is dismembered using one or more incisions and removed in parts. This is not without potential complications, including retained placenta, endometritis, peritonitis, lacerations of the birth canal, and acute laminitis. If fetotomy cannot be performed expeditiously with a minimum number of fetal incisions, it is safer to transport the mare to an equine hospital for the purpose of doing a cesarean section.

TREATING SPECIFIC DYSTOCIAS

It is important to be able to recognize specific dystocias as quickly as possible. Knowledgeable owners are in a better position to provide timely and critical information to their veterinarians.

Whenever possible, obstetric emergencies in mares should be treated by trained professionals. Veterinary practitioners have found by experience that well-intentioned but unnecessary interventions by those attending the mare's foaling often cause more problems than they cure. Exploratory vaginal examinations by untrained personnel, for example, seldom yield accurate information but are associated with the major risk of introducing infection into the reproductive tract.

Attempts to reposition and extract a foal can lead to tears of the cervix and vagina as well as rupture of the uterus. Pulling on the wrong part can wedge the foal more tightly into the pelvic canal and make the blockage worse. You could also break your arm if it gets caught between the foal and the rim of the pelvis during a forceful uterine contraction.

If you absolutely cannot get help and feel that intervention is unavoidable, keep in mind that unless you are skilled or fortunate enough to diagnose and treat a fetal emergency successfully in less than 20 minutes, the foal probably is not going to survive. The condition of the mare is not quite so urgent, but after about an hour's delay, the mare will begin to show signs of shock and require veterinary attention.

Most dystocias are easier to correct with the mare in the standing position. An assistant should restrain the mare with a lead and halter.

DEVIATION OF THE HEAD

A flexed head is the most common abnormal presentation in mares. In the ventral deviation, the head is tucked down against the chest with the nose wedged beneath the pelvic brim. The crown of the head and not the muzzle will be found between the feet.

In the lateral deviation, the neck is bent to the side with the nose pointing toward the foal's tail. This deviation is difficult to distinguish from *wry neck,* a congenital curvature of the cervical spine.

Either presentation can be suspected if you see both feet protruding from the birth canal without a nose between them. A mare of normal or larger size may be able to deliver a foal with a ventral deviation without assistance. She will not be able to deliver a foal with a lateral deviation.

With a ventral deviation, the muzzle must be disengaged from beneath the rim of the mare's pelvic bone before the head can be extended. Between contractions, one hand is used to push back on the crown or poll while the other is slipped beneath the foal's muzzle, lifting it up over the pelvic rim and into the vaginal canal.

When the neck is bent to the side, the situation is more difficult because it takes substantial strength to repulse the foal back far enough into the uterus to create space to reach in and straighten out the long neck. The mare is placed on her side with the head of the foal uppermost. While an assistant pushes back on both of the foal's front legs, the operator pushes on the neck. As the foal moves back into the uterus, the operator slides the other hand beneath the foal's chin, grasps the muzzle or lower jaw, and pulls the head foreword into the extended position for delivery. It might be necessary to expedite delivery by grasping the foal's front legs and applying moderate traction as the mare strains.

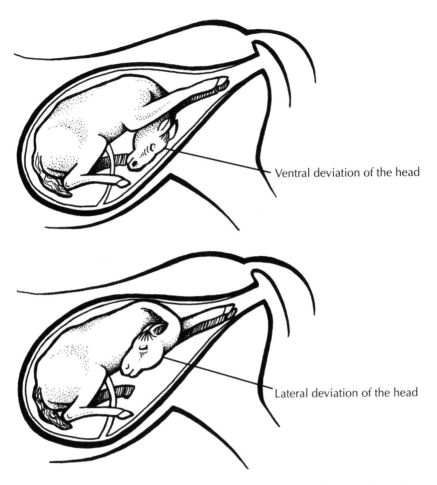

Deviations of the head (ventral and lateral) are common malpresentations. Suspect this if you see both feet protruding through the vulva without a nose between them.

DEVIATION OF THE FORELIMBS

Foreleg deviations constitute a second common group of malpresentations. In the *knee-flexion deviation*, one or both knees are tucked beneath the foal's chest. In the *shoulder-flexion deviation*, one or both forelegs lie alongside the foal's abdomen. The diagnosis is suspected by seeing the nose alone, or the nose along with one front foot, protruding through the vulva.

The exact position of the forelimbs (knee or shoulder flexed) can be determined only through vaginal examination by an experienced practitioner. When both forelimbs are deflected, it takes longer to correct the deviation. Speed is important.

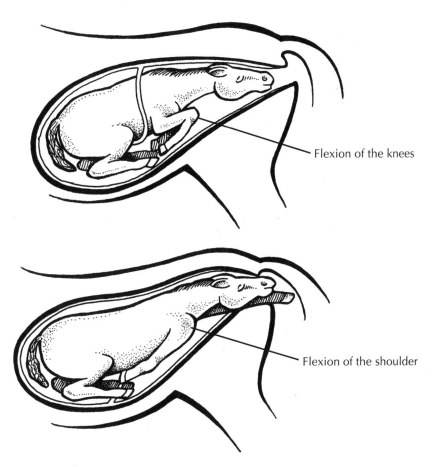

Deviations of the forelimbs, with knee-flexed and shoulder-flexed postures. Suspect one of these positions if you see a nose without both feet protruding through the vulva.

If the foal has one knee folded beneath him, the operator pushes against that knee and also the opposite, extended leg of the foal as the mare relaxes until he or she has room to slip a hand beneath the foal's knee. The operator then raises the foal's knee up toward the dome of the mare's uterus as far as it will go. This flexes the foal's elbow and shoulder. The operator then grasps the foal's leg close to his fetlock; as he straightens out the leg, he cups the foal's hoof in his hand to prevent injury to the mare's birth canal.

If the foal has both knees folded beneath him, the operator repulses the foal by pushing simultaneously on both knees. The above procedure is repeated on each side.

In the shoulder-flexion deviation, the entire head may be visible outside the birth canal. The mare's pelvic canal is lubricated as copiously as possible. The foal's head is pushed back into the mare's pelvis as far as it will go. The foal's foreleg is grasped above the knee, and the leg is drawn foreword into the bent-knee position. Then the bent-knee position is corrected as described above. If necessary, the procedure is repeated on the other side. Delivery must now proceed as quickly as possible. Traction is applied on the foal's legs as the mare strains.

FOAL ON ITS SIDE (DORSOILEAL POSITION)

As labor commences, rotational forces turn the foal from his back onto his side and eventually onto his chest and abdomen and into the dorsosacral position. If the rotational force is weak, the foal does not complete the full 180-degree rotation and enters the birth canal lying on his side. This abnormal position can be suspected by noting that the foal's feet point to the 3 or 9 o'clock position rather than to the 6 o'clock position. This finding can also occur with torsion of the uterus.

Unless the foal is small in relation to the diameter of the birth canal, he must be rotated into the dorsosacral position in order for delivery to continue. This is difficult to accomplish without obstetric chains and hooks. The birth canal is lubricated as copiously as possible. The operator slides one hand beneath the foal's lowermost shoulder, grasps the front leg near the shoulder joint, and lifts upward. Simultaneously, an assistant grasps the uppermost leg and pulls it down and in toward the midline. With several episodes of mare straining, the foal should rotate into the desired position. Moderate traction is applied to assist delivery.

Note: An arm can become seriously injured if trapped between the foal and the pelvis during a forceful contraction.

UPSIDE-DOWN FOAL (DORSOPUBIC POSITION)

Complete absence of rotation results in the foal coming out on his back. This abnormal presentation can be suspected by seeing the soles of the front feet opening pointing up to the sky. In addition, the muzzle will appear upside-down. In many cases, the foal is not exactly upside-down but is canted to one side or the other. In that case, the abnormal position is treated as described for the foal lying on his side.

If the upside-down foal is recognized early, the experienced practitioner can attempt to rotate the foal 180 degrees into the normal delivery position. If the foal is partway out, however, it is better to let the mare continue to deliver in this position. To prevent the foal's hooves from injuring the top of the birth canal, moderate traction is exerted on the foal's front legs, and the feet are held down until the shoulders clear the pelvis. If sufficient room is available, the foal's back legs and feet are also held down until they, too, pass through the pelvic outlet.

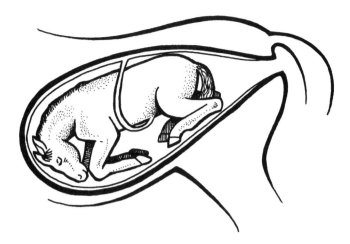

A posterior presentation. The leading parts are either the hind limbs extended (preferable), or as in this case the buttocks and tail (true breech).

POSTERIOR PRESENTATION

This rare presentation occurs only once in every 500 deliveries. The leading parts are either the hind limbs, or the buttocks and tail (true breech). The foal may be in the normal dorsosacral position or the upside-down dorsopubic position. One or both hind limbs may be extended (this is preferable), folded beneath the body in the hock-flexed posture, or stretched out beneath the body in the hip-flexed posture. In 90 percent of cases, the hock-flexed and hip-flexed postures are bilateral.

The preferred right-side up position with extended hind limbs can be suspected as the legs appear at the vaginal opening because they turn abruptly down. In the true breech, either no fetal parts or only the tail and rump are visible.

A well-recognized complication of the posterior presentation is compression of the umbilical cord between the foal's abdomen and the rim of the mare's pelvis. Delivery in this presentation must proceed rapidly to prevent fetal hypoxia.

When the foal's hind limbs are extended, each leg is grasped and forceful traction is applied in a downward arc as the mare strains. Tension between contractions is maintained. Be prepared to resuscitate the foal.

A true breech must be converted to a legs-extended posture and then delivered as above. This is a difficult maneuver. An upside-down or lateral position must be rotated into a dorsosacral position. Ordinarily this involves the use of chains, followed by forced extraction. A C-section is the treatment of choice if the foal is alive.

TRANSVERSE PRESENTATION

In this rare cause of dystocia, the foal lies crosswise in the uterus. As with a true breech, labor continues for 30 minutes or longer without any fetal parts becoming visible at the vaginal opening. On vaginal examination, either all four feet or the foal's back will be found in the birth canal. When multiple feet are present, the transverse presentation is often confused with a twin pregnancy.

The foal must be turned lengthwise in the pelvis. This is a difficult maneuver, even for experienced veterinarians. Prepare for a C-section.

FOOT-NAPE PRESENTATION

In this uncommon delivery, the foal's forelimbs lie on top of his head. The overriding position of the hooves presents a serious danger of rupturing the roof of the vagina and causing a rectovaginal fistula. On vaginal examination, the legs are equal in length and rest on top of the head.

The feet may need to be freed from lacerations in the vaginal roof before attempting to repulse the foal. An assistant raises the muzzle and pushes the head back into the pelvic canal as the mare relaxes. Simultaneously, the operator grasps both fetlocks and pulls down and out on each leg, bringing the forelimbs together underneath the head.

DOG-SITTING PRESENTATION (HIP-FLEXED POSTURE)

In this rare presentation, flexion of the hip joint causes the hind limbs to lie alongside the abdomen like a sitting dog, with the hooves engaging the wall of the uterus below the pelvic outlet. The forelimbs and head deliver normally, but as soon as the chest appears, progress comes to a halt.

The presentation is difficult to correct. The hind hooves must be disengaged from the mare's pelvis; then the hind limbs must be moved from hip flexion all the way back to hip extension. Forced extraction without correcting the abnormal leg posture is contraindicated because it drives the hind hooves through the wall of the uterus.

FETAL MALFORMATIONS

Approximately 35 percent of dystocias are caused by fetal deformities, the most common being contracted foals, hydrocephalus and fetal monsters.

The term *contracted foal* is used to describe a variety of spine and limb deformities that develop because of abnormal fetal position or congenital factors of unknown cause. The neck, spine and extremities are rigidly twisted into abnormal configurations. Many of these fetuses are aborted. If abortion does not occur and the mare is able to carry the foal to term, vaginal delivery is difficult or impossible owing to the flexed position of the forelimbs and increased diameter of the shoulders and chest. Either fetotomy or cesarean section will be necessary to deliver the foal. Some foals with back, neck and limb deformities born alive and able to nurse may recover spontaneously.

Hydrocephalus is enlargement of the bones of the skull caused by a buildup of fluid around the brain. The head is usually twice the normal size.

Vaginal delivery in the case of severe fetal deformities and monsters is usually impossible. The fetus must be removed by fetotomy or in some cases by cesarean section.

SMALL PELVIC OUTLET

A juvenile pelvis may be present in a young mare yet to attain full maturity. Another cause of a narrow birth canal is a previously fractured pelvis. A small pelvic outlet predisposes a foal to shoulder and hip lock.

If the dystocia cannot be corrected with obstetric manipulation, emergency cesarean section is indicated if the foal is alive. When a small pelvic outlet is recognized before labor, an elective C-section is the procedure of choice. The size of the pelvic opening can be determined during the breeding soundness examination.

TWINS

Spontaneous delivery of live twins is unusual. The presence of two fetuses prevents delivery because neither has enough room to enter the birth canal. The problem is further complicated when one or both of the foals are dead.

This situation is difficult to correct if it is not immediately recognized. When both twins are in a normal anterior presentation, it may be possible to repulse one back into the uterus in order to gain enough room to deliver the other. Once the first is delivered, the second follows rapidly. When twins are diagnosed before labor, an elective C-section is the treatment of choice.

HIP LOCK

The foal's hips can become engaged with the side wall of the pelvic ring much as the back wheel of a tricycle can lock on a doorjamb. Hip lock should be suspected when progress stops with the foal halfway out of the birth canal. The mare continues to strain but the foal does not move.

A well-recognized complication of hip lock is compression of the umbilical cord between the foal's abdomen and the mare's pubic bone. Delivery must proceed quickly to prevent fetal hypoxia. The hip is disengaged by grasping the foal's front legs and crossing the right leg over the left and twisting clockwise. This applies torque to the foal's body and pelvis. The foal is then pushed back into the uterus. If this does not disengage the hip, the left leg is crossed over the right and torque is applied in a counterclockwise direction. The foal is pushed back firmly. Some veterinarians find it easier to grasp both front legs and repeatedly pull the foal back and forth from left to right, wiggling the foal's pelvis through the mother's pelvis. Once the block is relieved, delivery proceeds rapidly.

ELBOW LOCK (SHOULDER-ELBOW FLEXION)

In this dystocia, one or both elbows bend and drop below the pelvic brim instead of remaining extended in the diving position. This creates a blockage and stops progress. If both elbows lock, the nose is visible at the level of the fetlock rather than at the level of the knee. If only one elbow locks, there is a greater discrepancy in leg lengths than the normal 3 to 6 inches.

This dystocia is relatively easy to correct because the forelimbs are already in the pelvic canal. The operator pushes the foal's head back into the birth canal with one hand, and with the other grasps the trailing hoof and pulls it foreword to straighten out the elbow. Birth should follow shortly.

SHOULDER LOCK

The width between the shoulders is the widest diameter of the foal. Dystocia can occur if both shoulders enter the pelvic ring at the same time. The blockage can be suspected by seeing the legs and head outside the birth canal without further progress despite abdominal straining.

Both legs are grasped, and the foal is pushed back into the uterus. Then one leg is pulled forward while the other is held back in the trailing position.

PREMATURE SEPARATION OF THE PLACENTA (PLACENTA PREVIA)

Separation of the placenta before the foal moves into the birth canal is rare. When this happens, the placenta precedes the fetus, and the water sac does not rupture. Signs are forceful abdominal straining and the appearance of a red-velvet, basketball-like structure (the placenta) at the vulva, instead of the bluish-white amnion. Behind the placenta is the intact water sac, which is prevented from entering the birth canal. The entire process arrests labor and must be immediately corrected to prevent fetal asphyxiation.

Treatment: If a veterinarian is not present, immediately rupture the protruding membrane using a pair of sharp scissors. This releases the contents of the water sac. Remove the placental membranes from the foal's nostrils. Labor should proceed quickly. Be prepared to revive a depressed foal. After the delivery, be sure to summon your veterinarian to examine the mare and the foal. Postpartum complications are common.

TORSION OF THE UTERUS

Torsion is rotation of the uterus about its long axis. It is an uncommon problem that occurs with equal frequency during late pregnancy and active labor. A torsion of 180 degrees twists the uterine birth canal and prevents the foal from emerging. A rare 360-degree rotation cuts off the blood supply to the uterus and results in death of the fetus within a matter of hours. The mare goes into shock and collapses.

The mare with a twisted uterus exhibits severe colic, but the pain is often attributed to labor. The diagnosis is made by rectal palpation that reveals a displacement of the broad ligaments. The direction of displacement shows the veterinarian in which direction the uterus has rotated.

A rotated uterus is best corrected by inserting a gloved arm through the cervix into the uterus. A firm part of the foal is grasped, and the foal and uterus are rocked back and forth until, with an extra effort in the direction opposite the torsion, the uterus is rotated back into its normal position. This maneuver is successful in about 80 percent of cases. In most of the remaining cases, the uterus can be replaced through an incision in the mare's flank (see PRETERM TORSION OF THE UTERUS in chapter 7). A C-section is indicated if the above methods fail or the foal is dead. Forced extraction results in rupture of the uterus and should not be attempted.

UTERINE INERTIA

Failure of the uterus to contract is called inertia. Inertia is classified as primary or secondary. Secondary inertia is caused by uterine muscle fatigue. The uterus strains to the point of exhaustion and is unable to expel the foal even after the blockage has been removed.

In primary inertia, labor proceeds slowly or stops because the uterine muscle does not respond appropriately to signals that initiate contractions. The muscle of the uterus is under the influence of oxytocin and prostaglandin. It has been suggested that a deficiency of one or both of these hormones may be the cause of some cases of delayed or prolonged labor. Activity of the foal is another stimulus to the uterus. A weak or depressed foal might not stretch the uterus enough to initiate powerful contractions. In older mares, the uterus may lack muscle mass because of age and chronic scarring.

A delay in labor for any reason is an urgent indication to summon your veterinarian. Treatment of secondary inertia is directed at relief of the blockage. If after relief the uterus is too weak to contract, your veterinarian may elect to give an injection of oxytocin. Oxytocin is also given in an attempt to initiate contraction in primary inertia. If uterine contractions cannot be initiated, the foal will have to be delivered by cesarean section.

CESAREAN SECTION

A cesarean section is used to retrieve a live foal when vaginal delivery cannot be accomplished. It is also indicated after rupture of the uterus, to correct torsion of the uterus, and to remove a dead fetus that cannot be removed by fetotomy.

Elective and emergency C-sections are done at equine veterinary hospitals and at stud farms with operating room facilities. Emergency C-sections in the stable or field—away from anesthesia, padded operating tables, sterile instruments and support facilities—cannot be done successfully.

When a C-section is done electively or shortly after recognition of a dystocia, the survival rates are 80 percent for mares and 30 percent for foals. One reason for the low survival rate in foals is the 33 percent occurrence of fetal deformities that occur among mares requiring C-section.

Survival rates diminish rapidly with the length of labor and the time it takes to proceed to surgery. When C-section is used as a last resort, the prognosis for the foal and the mare is poor.

The operation can be done through the flank or abdomen. The abdominal approach is preferred by most veterinary surgeons. The mare is given a general anesthetic and placed on her back. A midline incision is made. The foal is removed, and the incision in the uterus is closed with interlocking sutures to prevent postoperative bleeding. If the placenta does not separate at the time of surgery, it can be left in place to pass at a later time. Antibiotics, anti-inflammatory drugs and oxytocin are administered during and after surgery.

Complications in the mare include postoperative bleeding from the uterus, peritonitis, retained placenta, delayed uterine involution, endometritis and peritoneal adhesions. Scarring of the uterus greatly reduces future fertility.

A mare who has had a C-section may not recognize her foal or be willing to accept it. Occasionally this can be overcome, as described in chapter 13 (see ABNORMAL MATERNAL BEHAVIOR). If the mare steadily refuses to accept her foal, the infant should be raised as an orphan.

CARE OF THE NEWBORN FOAL

The newborn or neonatal period is defined as the interval from birth to 1 month of age.

AFTER THE DELIVERY

NURSING BEHAVIOR

During the first week of life, a newborn foal nurses every 15 to 20 minutes. The most important indicator of how well the foal is doing is how well he is nursing. A foal who is sick or not adapting well for any reason manifests weakness and depression by consuming less milk. One of the most reliable indicators of nursing efficiency is the appearance of the mare's udder. Healthy, vigorous foals keep their mothers drained of milk at all times. An udder that remains full and distended is not being utilized.

A foal with a weak suckling reflex who is not able to swallow the milk ejected from the mammary gland in response to his efforts at suckling may have milk on his face. Any foal who is nursing apathetically should be considered at risk for septicemia. Contact your veterinarian.

The healthy foal nurses vigorously and often.

MECONIUM COLIC

A normal newborn foal passes meconium without straining in the first 3 hours after birth. Meconium is a greenish-brown to black, hard, stool-like material that accumulates in the foal's intestinal tract before birth. By the 4th day, it is replaced by the yellow feces of the nursing foal.

When meconium is not passed within 6 to 12 hours, it becomes painfully impacted in the rectum or colon. Signs of distress appear within 24 hours of birth. The foal nurses less frequently, arches his back and strains without passing stool, moves about the stall with his tail elevated, swishes his tail, repeatedly gets up and down, and exhibits colic with rolling and thrashing. The diagnosis is established by digital rectal examination which reveals firm meconium in the rectum.

Treatment involves softening up the hard meconium with enemas. A prepackaged Fleet™ phosphate enema usually relieves an impaction limited to the rectum. While an assistant restrains the foal, the tip of the tube is lubricated and inserted its full length into the foal's rectum; the enema is administered without excessive pressure. Hold the tail down for a few minutes to prevent the immediate return of the enema. If the impaction is not relieved within 1 hour, follow with a soapy water enema. A second Fleet enema is permissible, but only if most of the first enema has been expelled. Fleet solution accumulating in the colon has been known to cause phosphate intoxication.

To prepare a soapy water enema, add 1 or 2 drops of Liquid Ivory™ soap (not detergent) to 1 to 2 pints of lukewarm water. The water should become milky. Fill an enema bag (or an empty Fleet bottle) with the solution. Connect the enema

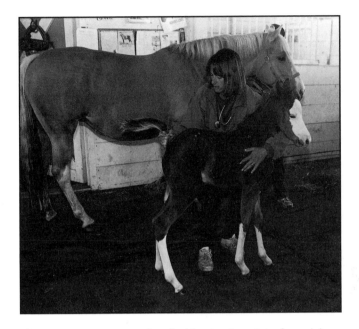

The correct way to restrain a foal for treatment, such as giving an enema.

bag or Fleet bottle to a soft, flexible, rubber catheter. Lubricate the catheter with mineral oil, and gently insert it 3 to 4 inches into the foal's rectum. Do not force the catheter against resistance, because this can perforate the rectum. Administer the enema slowly. Do not give more than 2 pints of soapy water.

For an impaction involving the large intestine, your veterinarian may administer an oral dose of mineral oil (8 ounces), and/or a laxative such as milk of magnesia (8 ounces). Mineral oil must be administered by a stomach tube because of the danger of aspiration pneumonia.

A small number of foals fail to respond to enemas and laxatives and develop a bowel obstruction. These foals require surgery.

HEALTH CARE

The newborn foal is capable of forming antibodies and can be vaccinated against rhinopneumonitis and influenza at birth if the mother was not vaccinated during late pregnancy, and if the prevalence of local disease warrants early protection. The immune response, however, will not be as strong as it will be when the foal is 3 months old. Accordingly, foals vaccinated at birth should be revaccinated at 2 months of age. Then follow the immunization schedule recommended in the appendix.

Tetanus antitoxin has been implicated as a cause of serum hepatitis and is no longer routinely administered to newborn foals (see TETANUS later in this chapter).

Prophylactic antibiotics have not been shown to reduce the occurrence of neonatal infections and may be associated with foal diarrhea. Accordingly, they are no longer advocated.

EXERCISE

Exercise is beneficial for both the mare and the foal. They can be turned into an empty paddock or a small field for 15 to 30 minutes on the second postpartum day. Increase the exercise so that after 2 weeks (weather permitting), they can be left outside most of the day. Do not turn them out with other horses until the foal is at least 3 weeks old.

Exercise is good for the foal and mare. By 2 weeks, they can be left outside most of the day. Photo courtesy Don Lupton.

PROBLEMS WITH LACTATION

The mare's udder is composed of two separate milk sacs that lie on either side of the midline between her back legs. Each milk sac is composed of two mammary glands. The ducts of the two milk glands come together to form a common duct that opens at each nipple or teat.

Milk letdown is controlled by oxytocin released from the pituitary gland during labor and delivery. Tactile stimulation of the teats (suckling, washing the teats) helps to trigger the release. Milk released from the mammary glands is stored in the cisterns that empty into the teats.

FAILURE OF PASSIVE TRANSFER (FPT) OF MATERNAL ANTIBODIES

Late in pregnancy, the mare produces a special milk that is high in fat, vitamins, minerals and protein. This is the colostrum, or first milk of the dam. Of singular importance, colostrum contains the immunoglobulin antibodies and other immune substances (primarily IgG, but also IgM) that protect the foal from neonatal infections. Failure of passive transfer (FPT) of immunoglobulins from dam to foal has been cited as the single most important cause of death during the first week of life.

In many species, including man, the fetus receives immunity in utero through the passage of antibodies across the placenta. This does not happen in the horse. The foal is born without circulating immunoglobulins and can acquire them only through the ingestion of colostrum during the first few hours after birth. Through a special adaptation in the cells of the intestinal lining, maternal antibodies are allowed to cross the mucosal barrier and enter the bloodstream. The capacity to absorb antibodies is greatest during the first 8 to 12 hours. Both the production of colostrum and the absorption of it by the foal taper off rapidly thereafter. After 24 hours, this barrier closes and antibodies are no longer absorbed.

FPT can occur for a number of reasons, including death of the dam, premature lactation prior to delivery, failure to secrete immunoglobulin into the colostrum, delay in suckling due to foal weakness or illness, and maternal rejection of the foal. If the mare was not immunized against the common infectious diseases, her colostrum will not provide passive immunity against those diseases.

Premature lactation (referred to as galactorrhea) is lactation prior to delivery. When it occurs in the third trimester, it usually indicates a pregnancy problem such as impending abortion, death of a twin fetus in utero, premature separation of the placenta, or placental infection. Premature lactation, even when it occurs a few hours before foaling, is a serious problem because it significantly lowers the IgG content of the milk suckled by the foal. Any mare who drips milk prior to foaling should be milked out so that her milk can be frozen and given to the foal after delivery.

The IgG content of the colostrum can be evaluated on the basis of its specific gravity. This is done by collecting a small amount of colostrum and placing it in a modified hydrometer (Equine Colostrometer; Lane Manufacturing Inc.; Denver, Colorado) similar to the instrument used in hospitals to measure the specific gravity of urine. Good-quality colostrum should have a specific gravity of at least 1.060. (By convention, the specific gravity of distilled water is 1.000.) If the specific gravity is less than 1.060, the antibody content of the colostrum may be too low to provide adequate immunity. Any mare who lactates prematurely should have her milk analyzed before the foal begins to suckle. This will alert her attendants to the possible need for obtaining supplemental colostrum.

The concentration of IgG in the foal's blood can be determined by a rapid stallside test kit (zinc sulfate turbidity, latex agglutination, *ELISA*, or hemagglutination inhibition test). The minimum concentration required for adequate protection is considered to be 400 mg/dL, *assuming that the foal is not infected*. If the foal is sick or at increased risk of infection because of unsanitary surroundings, the minimum concentration should be 800 mg/dL.

Treatment of IgG Deficiency: Replacement therapy should be initiated immediately on the diagnosis of IgG deficiency. Many breeding farms maintain a colostrum bank, obtained by milking 4 to 6 ounces of colostrum from mares immediately after foaling and then pooling it and freezing it for later use. This amount is safe to collect without depriving the foal.

A total of 2 to 3 liters of colostrum divided into three to four doses should be given as soon as possible within the first 12 hours after birth. The colostrum can be given by a nursing bottle or preferably by a stomach tube.

Lyophilized (freeze-dried) equine IgG (commercially available as Foal-Aide from Equilab, Rockville, MD) can be given as a colostrum substitute. Up to 60 grams orally are needed to raise plasma IgG concentrations from 0 to above 400 mg/dL. Pooled equine plasma can also be given orally, although large volumes (6 to 9 liters) are required to achieve adequate immunity.

At 18 to 24 hours after colostrum replacement, the foal's serum immunoglobulin concentration should be checked to determine if protection is adequate.

Frozen bovine (cow) colostrum can be substituted for mare's colostrum. The half-life is shorter and the antibody coverage is somewhat different from that needed for foals. The principal advantage of bovine colostrum is that it is more readily available than mare's colostrum. Note that equine IgG kits cannot be used to measure the concentration of bovine IgG in the serum of foals.

If more than 18 to 24 hours have elapsed without colostrum replacement, colostrum and colostrum substitutes must be given by the intravenous route. Foals with serum IgG levels below 400 mg/dL should be given 1 to 4 liters of pooled equine plasma by vein. The exact amount depends on the level of circulating IgG before treatment, as well as the potential for foal septicemia. Equine plasma can be purchased (Equine Plasma; Veterinary Dynamics, Inc.; Chino, California) or drawn from local horses by your veterinarian.

Following treatment, the serum IgG level should be checked to determine if adequate amounts were given.

Newborn foals are dependent on colostrum to provide vitamin A. The colostrum-deficient foal should be given a prophylactic vitamin A injection.

AGALACTIA (INSUFFICIENT MILK SUPPLY)

For the first 2 to 3 months, the foal's nutritional requirements are met entirely by his mother's milk, even though he begins to nibble on the mare's grain and hay when only a few days old.

Absence of milk is called agalactia. Failure to lactate immediately after foaling is highly significant because of the problems associated with FPT (discussed above). Mares on tall fescue pastures in the Southeastern United States experience a high incidence of agalactia due to ingestion of a specific fungus that grows on the grass. The fungus produces an alkaloid that suppresses the pituitary hormone prolactin, which is responsible for initiating lactation.

Some mares are poor milk producers. They do produce some milk, but not enough to satisfy their foals. A foal who is not getting enough milk is thin, nurses vigorously, and never seems to be satisfied.

One cause of insufficient milk production is a diet low in energy, protein and essential nutrients. Initially the mare who is not getting enough nutrition will make up the difference at the expense of her body stores. It is only after her weight drops to a body-condition score of 4 or less that her milk production begins to decline. Inadequate milk production caused by poor nutrition can be prevented by feeding an appropriate diet, as described in FEEDING FOR LACTATION in chapter 13.

Selenium deficiency and an inherited failure of the mammary glands to produce milk have also been cited as causes of agalactia. There is a type of mastitis caused by eating the leaves, bark or fruit of the avocado tree. In avocado poisoning, milk production stops and does not resume.

Treatment: When milk fails to come in spontaneously after foaling, the application of warm compresses to the udder and an injection of oxytocin may stimulate milk letdown.

Mares with fescue-induced agalactia have been successfully treated with an experimental drug called domperidone. Affected mares with little or no udder development prior to foaling who received domperidone for 10 to 14 days developed normal udder size within 5 days of treatment. The newborn foals of these mares were able to nurse effectively and receive healthy colostrum. Domperidone also has been effective in treating other causes of agalactia, including insufficient milk production—thus eliminating the need to acquire milk from some other source.

When milk production cannot be initiated by one of the above treatments, the foal should be treated for IgG deficiency and raised as described below in THE ORPHAN FOAL.

MASTITIS

Inflammation of one or more quarters of the mammary gland is called mastitis. Mastitis is rare in mares.

Mastitis usually occurs several weeks after foaling but may occur after weaning and even in mares who have never been pregnant. The signs are a warm, swollen, painful udder; mild hind-limb lameness; loss of appetite and depression. The udder swelling may extend to involve the undersurface of the abdomen. The mare often refuses to nurse. Milk from the infected gland appears curdled and contains blood. Laboratory examination of the milk will show a large number of PMNs and bacteria. Culture reports usually reveal streptococci or staphylococci.

Treatment: The foal should be removed temporarily and fed by milk bucket or nursing bottle, as described below in THE ORPHAN FOAL. Milk out both teats to relieve pressure and perpetuate milk production. Cold packs help to reduce the swelling. Antibiotics are started and adjusted according to the results of culture and sensitivity tests. The infection usually responds promptly, and the foal can return to nursing in 1 week. If milk production is allowed to stop, however, it might not resume when the foal returns to the udder.

UDDER EDEMA

A large amount of *edema* fluid may accumulate in the lower abdominal wall and udders before or after foaling. This is most common in *primiparous* mares. The swollen udders may produce enough distress so that nursing is painful and may be resisted.

Treatment: The swelling disappears spontaneously. Treatment, which is indicated only if edema interferes with nursing, involves the use of ice packs, hydrotherapy, light exercise, and occasionally the administration of Lasix™.

THE ORPHAN FOAL

A foal may be orphaned because of the death of his mother or because the mother is unable or unwilling to nurture the foal for reasons of ill health or lack of maternal bonding. If the foal did not receive colostrum within the first 8 to 12 hours, follow the procedure described in FAILURE OF PASSIVE TRANSFER (FPT) OF MATERNAL ANTIBODIES above.

An orphan foal can often be raised by another lactating mare. Most mares will accept a substitute foal within 12 to 72 hours of foaling but usually must be restrained until the foal establishes a maternal bond through repeated nursing. The mare should not be tranquilized. Most tranquilizers are transferred in the milk and can sedate the foal.

Mares rely primarily on the sense of smell for infant recognition. It is helpful to make the foal smell like its new mother. You can smear the foal with amniotic fluid or coat the foal with the mare's sweat, milk or feces.

On some farms, the use of a milk goat has been successful. Many nannies readily accept foals. The composition of goat's milk is different from mare's milk, but not enough to cause problems. The nanny (or nannies) will have to stand on a raised platform or a bale of hay for the foal to nurse; or the foal will need to nurse on its knees. Even if the nanny does not work out as a foster mother, nannies provide good companionship.

If a foster mother is not available, the foal can be raised by hand. This involves the use of a milk substitute or milk from another source, attention to feeding, and the hygienic care of all nursing equipment.

A good-quality commercial milk replacer such as Foal-Lac™, Mare's Match™ or Mare Replacer™ is the most balanced substitute for mare's milk, but milk replacers are not as palatable as mare's milk and are also quite expensive. Again, goat's milk can be considered since it satisfies the basic nutritional requirements and may be readily available from a nearby goat dairy. Goat's milk can be fed full strength or diluted half and half with milk replacer. If none of these are available, you can use a commercial calf-milk replacer on a temporary basis.

The dietary energy and water requirements of the young foal are extremely high. A large volume of milk must be consumed to meet these needs. This amounts to 14 quarts per day for the first week. You then add 1 quart for each additional week of age until the foal is consuming 18 to 20 quarts per day. Milk replacer is reconstituted with water according to the directions provided. Follow the manufacturer's instructions in regard to frequency and volume to feed. Most milk formulas are fed in four or more equal feedings, beginning with one-half the recommended daily requirement during the first 24 hours. The reason for starting slowly is to avoid diarrhea.

The milk formula can be given by bottle or bucket. Bucket feeding is easier and safer than bottle feeding. Start with a pan shallow enough so that the foal's nose can touch the bottom without being submerged. Dip your fingers into the pan and allow the foal to suck on your fingers. If the foal does not suck well, move your fingers against his tongue and palate to stimulate a better suckling reflex. Once the foal begins to suck, encourage him to drink from the pan by lowering your fingers into the pan. It can take up to 2 hours to teach a foal to drink from a pan, but with time and patience, it is not difficult to accomplish.

Once the foal learns to drink from the pan without submerging his nostrils, switch to a plastic bucket with a wide opening. Hang the bucket in a convenient location and keep it full at all times. Keep in mind that foals drink small amounts at frequent intervals during both day and night. Clean all feeding equipment and change the bucket every 12 hours. In addition, provide continuous access to fresh water.

Foals do not take naturally to bottle feeding. Initially, the milk tends to run out the mouth and nose. In addition, if the foal is fed with his head held up, a certain amount of formula will reflux into his breathing passages. To avoid aspiration pneumonia, it is essential that the bottle be held down low so that the foal's muzzle is below the level of his larynx.

When the foal is 1 week old, begin feeding milk-replacer pellets available from feed stores. Start by putting the pellets in the foal's mouth several times a day. Then put as many pellets in a bucket or creep feeder as you think the foal will eat, and keep it filled. Discard the old pellets twice a day. Also at this time, provide high-quality hay free-choice for the foal to nibble.

After the foal is eating 2 to 3 pounds of milk-replacer pellets a day, add a high-quality creep-feed ration. A creep-feed *ration* is a mix composed of processed grain specifically formulated to meet the needs of the nursing foal. The National Research Council recommends that a creep feed contain 16 percent crude protein,

0.9 percent calcium and 0.6 percent phosphorus, based on total weight of the ration. Appropriate creep feeds can be purchased from feed stores or formulated using grains and concentrates. When the foal is 3 month old and eating well, the milk feedings can be stopped.

As soon as the foal is eating 4 to 6 pounds of the creep feed/pelleted mixture a day, the pellets can be stopped. At 4 months of age, the orphan foal can be fed as a weanling. For information on feeding weanlings and yearlings, see *Horse Owner's Veterinary Handbook* (Howell Book House, Inc.).

NEONATAL DIARRHEA

Diarrhea is the most frequent problem affecting newborn foals. Diarrhea has many causes, some of which are mild and inconsequential. Infectious diarrhea, however, is serious and potentially fatal. Familiarity with neonatal diarrhea will help you to decide when to seek professional assistance.

FOAL HEAT DIARRHEA

Foal heat or "9th day diarrhea" (which occurs from day 6 to 14) affects nearly all newborn foals. The diarrhea is mild and is over in less than 7 days. The manure is soft, pasty-yellow, and not profuse. The foal remains bright and alert and nurses at regular intervals. The unaffected behavior of the foal distinguishes foal heat diarrhea from more serious diarrhea.

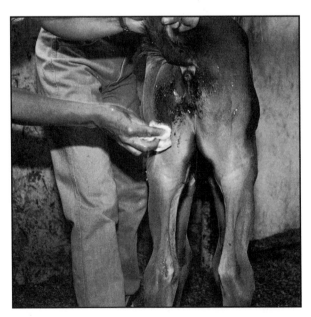

Foal heat diarrhea. Keep the perineum clean and apply Vaseline to prevent scalding.

Foal heat diarrhea is only incidentally related to the mare's estrous cycle and has nothing to do with the concentration of hormones in her milk. Newborns normally eat manure and feedstuffs such as grain and hay. It appears that the ingestion of these substances may upset the flora of the foal's immature intestinal tract and cause the temporary diarrhea.

Treatment: Diarrhea of short duration associated with the foal heat requires little treatment. Keep the foal dry and clean around the tail and perineum. Apply zinc oxide ointment or Vaseline to prevent scalding. Do not use laxatives or intestinal purgative, as this makes the diarrhea worse.

NUTRITIONAL DIARRHEA

A common type of mild diarrhea is associated with the consumption of more milk than the foal can absorb. It occurs in foals whose mothers are heavy milk producers.

Additionally, a temporary lactase deficiency can result in carbohydrate intolerance and subsequent diarrhea. This can occur in neonates who are either being hand fed or are recovering from viral enteritis. See ROTAVIRUS INFECTION (VIRAL ENTERITIS) below.

Treatment: Diarrhea caused by excessive milk consumption can be helped by milking out the mare 2 or 3 times a day. Nursing should not be restricted, because the milk is the foal's only source of water. As the foal gets larger, his nutritional needs increase accordingly and his intestinal tract will absorb more milk. This corrects the problem. A temporary lactase deficiency can be treated by giving the foal an oral lactase preparation such as LactAid,™ available at health food stores.

PARASITIC DIARRHEA

Threadworms are transmitted in the mother's milk and are the first intestinal parasites to mature in the foal. They are an infrequent cause of diarrhea. A large burden of worms is required to cause symptoms. Threadworm diarrhea can appear during the first 2 weeks of life. Thus it may overlap foal heat diarrhea.

Diarrhea caused by large strongyles and ascarids usually affects foals older than 1 month of age.

Treatment: Ivermectin is effective against threadworms, ascarids, and large and small strongyles. On farms where threadworms have been a problem, mares should be given ivermectin or oxibendazole within 24 hours of foaling to prevent the passage of threadworms in their milk.

BACTERIAL DIARRHEA

Several bacterial species produce a severe and often fatal infection of the intestinal tract of newborn foals. Lack of colostrum and failure of passive transfer of maternal antibodies is involved in a high percentage of cases.

Salmonellosis. Salmonella infection is a leading cause of newborn septicemia. This bacteria enters through the intestinal tract and often settles in other body systems, including the lungs, joints, eyes, kidneys and brain.

Several species of salmonella cause infectious *enteritis* in foals 1 to 4 months of age. The signs are fever, weakness, and a profuse, watery diarrhea that may contain blood. The foal becomes depressed, stops suckling, and quickly dehydrates.

Clostridial Infection. Both *C. perfringens* and *C. difficile* produce highly lethal infectious diarrhea in newborn foals. The diarrhea is bloody, foul-smelling, profuse, and associated with severe colic. It is unusual for a neonatal foal with clostridial diarrhea to survive more than 48 hours. In fact, the foal may be found dead before the diarrhea develops. Clostridia are normal inhabitants of the horse's colon; therefore recovery of the bacteria does not necessarily prove a causal relationship. Finding the endotoxin in the foal's stool or serum is more conclusive.

Vaccination of pregnant mares with *C. perfringens* toxoid has been advocated on farms with a history of clostridial infections in foals. The safety and effectiveness of this vaccine is not yet established.

E. Coli Infection. E. coli is the most common bacteria isolated from blood cultures in septicemic foals. This bacteria tends to seed in multiple organs. Diarrhea is a late and often terminal event.

Treatment: All foals with diarrhea who appear ill, depressed, and/or stop nursing should have an immediate veterinary examination. The diagnosis of bacterial enteritis is based on laboratory studies including blood and stool cultures. While awaiting results, the foal is started on a broad-spectrum antibiotic that can be changed later as culture and bacterial sensitivity reports become available.

For the foal with diarrhea who is mildly dehydrated and continues to nurse, dehydration can be corrected with oral electrolyte solutions administered by stomach tube. The severely dehydrated foal is given fluids by the intravenous route. Plasma is given to replace protein losses. Salmonella species rapidly develop a resistance to antibiotics. This renders antibiotics less effective for salmonellosis than for other types of bacterial enteritis. However, antibiotics are given in all septicemic diarrheas to control symptoms and prevent the bacteria from invading the joints, lungs and other organs. Supportive treatment as described in FOAL SEPTICEMIA later in this chapter is of great value.

ROTAVIRUS INFECTION (VIRAL ENTERITIS)

Rotavirus is a highly contagious diarrhea that attacks foals up to 6 months of age. Most cases occur in 2-month-olds. This correlates with the interval during which the foal's maternal antibody levels are in natural decline. While neonatal infection is uncommon, the consequences are serious because dehydration occurs rapidly.

The illness begins with pronounced apathy and a high fever. Some foals experience a mild infection, pass manure of cowpie consistency, nurse lethargically for 1

Skin ulcerations associated with the watery diarrhea of a rotavirus infection.

or 2 days and then recover. In others, the diarrhea is explosive, appearing watery green to gray. The acute form can last 5 to 7 days and result in significant fluid and electrolyte losses.

During the acute infection, the rotavirus destroys the intestinal brush border cells lining the upper intestinal tract. These are the cells that produce lactase. Their destruction results in lactase deficiency. Regeneration can occur within a few days, but some foals remain lactase deficient and are unable to absorb nutrients for some time. This is the cause of a nutritional diarrhea that persists long after the rotavirus is eliminated.

Treatment: Administer Pepto-Bismol™ (20 ml per 100 lb body weight) by syringe or tablespoon three to four times a day to control diarrhea and protect the lining of the intestine. Restrain the foal and insert the medicine into the corner of his mouth. Zinc oxide ointment or Vaseline™ is recommended for skin scalds involving the perineum and hind legs. A severely dehydrated foal requires IV fluids and hospitalization. Diarrhea associated with a lactase deficiency is managed by switching to solid feeds and weaning the foal as early as possible

Prevention: The rotavirus is rapidly transmitted through contact with contaminated feed, water, bedding, human hands, grooming utensils and other sources. Sick foals should be isolated to prevent the spread of infection. Thoroughly disinfect the premises and all equipment to destroy the virus. Formaldehyde (not bleach) is the disinfectant of choice.

A vaccine for pregnant mares is under investigation. Preliminary studies indicate that the vaccine induces antibodies that are at least partially protective for the foal during the first 3 months. The vaccine is not licensed for general use. Horse owners considering the use of equine rotavirus vaccine should obtain permission for its use from state veterinary authorities.

GASTRIC AND DUODENAL ULCERS

Ulcer disease is common in foals. It has been found that up to 50 percent of foals develop ulcers, the majority during the first 4 weeks of life. Ulcers occur either in the stomach or just beyond the outlet of the stomach in the small intestine (duodenum).

Stress appears to play a major role in causing ulcers in foals. Stress affects foals in the neonatal period in association with diarrhea and other neonatal illnesses. It also occurs at 2 to 3 months of age when maternally acquired immunity begins to wane, and again at 4 to 5 months in association with weaning.

Most ulcers do not produce symptoms and disappear with time. When symptoms do occur, they include abdominal pain, grinding of the teeth (called bruxism), frothy salivation, poor appetite, diarrhea of varying frequency, poor growth, rough hair coat, a potbellied appearance, and a tendency to lie on the back.

Symptomatic disease is more likely to occur when ulcers have produced scarring in the stomach or duodenum. This causes a stricture that partially or completely prevents the stomach from emptying. These foals may reflux stomach contents up into their throats and develop aspiration pneumonia.

One other complication is rupture. Intestinal contents spill into the abdomen and cause a fatal peritonitis. The diagnosis can be confirmed by *abdominocentesis* with the finding of infected fluid and food particles in the peritoneal cavity.

Examination of the interior of the stomach by fiberoptic endoscope (*gastroscopy*) is a convenient way to make the diagnosis and determine the number, location and size of the ulcers. X-ray studies using barium are helpful in showing an obstructed stomach.

Treatment: It is important to treat coexistent illnesses and remove all causes of stress. Anti-ulcer drugs such as the hydrogen ion blocker cimetidine, along with antacids, are the mainstays of treatment. Your veterinarian will determine the selection, dosage and frequency of administration of anti-ulcer drugs.

Foals with gastric-outlet obstructions that fail to respond to intensive medical management should be treated surgically. A ruptured ulcer with peritonitis cannot be treated.

NEONATAL INFECTIONS

FOAL SEPTICEMIA

Foal septicemia is a rapidly progressive bacterial infection of the bloodstream that affects foals in the first week of life. It is the most common cause of severe illness and death in newborn foals. Foal septicemia is a spectrum of infections caused by a number of different pathogens. Some, like actinobacillosis, are recognized as specific diseases.

Any foal who appears noticeably lethargic and does not stand within 1 hour or suckle within 3 hours is at increased risk of septicemia or may already be infected.

The most important predisposing factor in the development of foal septicemia is lack of colostrum with failure of passive transfer of maternal antibodies (discussed earlier in this chapter). Other predisposing factors include prematurity; a complicated birthing; overcrowded, poorly ventilated surroundings; failure to use good hygienic birthing practices such as wrapping the mare's tail and washing the udder and perineum; and failure to disinfect the umbilical cord stump with 2 percent iodine.

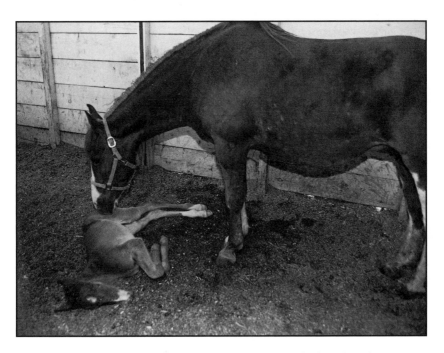

A lethargic, recumbent foal at 2 days of age, suggesting foal septicemia.

Bacteria gain entrance to the body through a variety of sites. The respiratory and gastrointestinal routes appear to account for the majority of cases. Some infections are acquired in utero while others occur during passage through the birth canal. Mother's milk may contain pathogenic organisms. The infected umbilical stump is a well-recognized cause of septicemia.

Signs of foal septicemia generally appear at 2 to 4 days of age and are followed by rapid deterioration. Early signs are weakness, lethargy, and reluctance to nurse (often indicated by engorgement of the mare's udder). An indication of lethargy is an unusual amount of time spent sleeping or lying on the side. Coughing or diarrhea may be the presenting symptoms. By the time the foal develops difficulty breathing, swollen joints, shock and collapse, the outlook for recovery is poor.

Treatment: Broad-spectrum antibiotics given IM or IV should be started on suspicion of foal septicemia, even though laboratory data and blood culture reports may not be back. After the reports become available, treatment can be modified if necessary.

A serum immunoglobulin will show whether the foal has adequate maternal antibodies. If the IgG concentration is low, the foal can be transfused with 2 to 4 liters of hyperimmune plasma.

Supportive therapy is absolutely necessary and often makes the difference between survival and death. Vigorous fluid replacement is essential to prevent and treat dehydration and shock. Recumbent foals should be kept on a soft mattress or suitable protective surface and maintained in the sternal position. If the foal cannot be maintained in the sternal position, turn the foal from side to side every 2 hours. Prevent chilling and hypothermia with blankets, heating pads and radiant heat lamps. Intranasal oxygen helps foals with respiratory difficulty. Sterile eye lubricants should be applied several times a day to prevent corneal injury caused by drying of the surface of the eyeball. The perineum must be kept clean and dry. Apply zinc oxide ointment to prevent skin scalds caused by liquid feces. The umbilical stump should be cleaned twice daily and disinfected each time with iodine.

Nutritional support is critical. If the foal is unable to stand, hand milk the mare and give the milk to the foal by stomach tube. Mare milk replacer can be substituted for mother's milk if necessary.

Following a septicemic episode, a foal may develop localized infections in various body cavities. Joint ill is a common sequel, as is *panophthalmitis*, an inflammation of all the inner structures of the eyes. Infection of the surface of the brain (*meningitis*) is another sequel to foal septicemia. Foals with meningitis develop seizures and often become comatose.

ACTINOBACILLOSIS (SLEEPY FOAL SYNDROME)

The bacteria *Actinobacillus equuli* is a common cause of foal septicemia. Sudden death can occur as early as 6 hours after birth. The bacteria may be present in the reproductive system of seemingly healthy mares without producing illness, which may explain why some foals are infected at birth.

Infected foals are extremely weak and have difficulty standing and nursing. They appear to sleep all the time and may be found in a coma. Diarrhea is common. Foals who survive often develop joint ill. Treatment is the same as that described for foal septicemia.

NAVEL ILL (UMBILICAL STUMP INFECTION)

Navel ill is a bacterial infection arising from the stump of the umbilical cord. The structures in the stump include the umbilical artery and vein, and the urachus. The infection is acquired either during foaling or shortly thereafter from contamination of the healing stump. Cutting or ligating the cord predisposes the foal to stump infection. A number of bacterial species can be involved; *streptococci* are the most common.

An infected umbilical stump is usually warm, tender and swollen (but may appear normal). A pus-like discharge is often present. The foal appears apathetic and stops nursing. In the septicemic form, bacteria enter the bloodstream through the umbilical vessels and spread to the liver and elsewhere. The foal becomes markedly depressed, goes into shock and dies in 12 to 24 hours. Some foals do not develop septicemia but instead develop signs of joint ill.

An ultrasound exam of the umbilical structures helps to make the diagnosis in atypical cases and can determine whether an abscess is present beneath the umbilical ring.

Treatment: Intravenous antibiotics are started at the first indication of navel infection. An abscess should be opened and drained and the infected umbilical structures surgically removed. Supportive care is important. Keep the umbilicus clean and dry, and apply a topical antiseptic such as Betadine solution twice daily. Change bedding frequently and keep the stall as clean as possible. Foals who develop septicemia are treated as described in FOAL SEPTICEMIA.

Navel infection can be prevented by good foaling practices, such as providing sanitary quarters and applying a 2 percent iodine solution to the umbilical stump after the foal is born.

JOINT ILL (SEPTIC ARTHRITIS AND OSTEOMYELITIS)

In this common sequel to foal septicemia, bacteria from the bloodstream invade the joints. Any neonatal infection can produce joint ill. Navel infection is the predisposing cause in 25 percent of cases.

The signs are sudden lameness, along with the appearance of one or more hot, swollen joints. The diagnosis is confirmed by inserting a needle into the joint and removing fluid for analysis. Fluid also is submitted for bacterial culture and sensitivity testing. Ultrasonography is useful in distinguishing between an abscess around a joint and infected fluid within a joint. Serial X-rays are necessary to identify secondary infection of the bone (osteomyelitis).

Treatment: Treatment is most effective when started shortly after the onset of lameness and swelling. Antibiotics in high concentrations are given for 4 to 8 weeks. Infected joints are irrigated daily for several days, using large-bore needles to infuse and remove the fluid. When the infection has been present for more than 1 week, it may be necessary to open a joint to remove pus and debris. A suction drain is inserted and the leg is immobilized beneath an occlusive dressing. Stall rest and physical therapy are important to a successful outcome. The outlook depends on the number of infected joints and their locations. Foals with secondary bone infection usually develop permanent arthritis.

FOAL PNEUMONIA

Foal pneumonia is a serious and often fatal infection that targets newborn and weanling foals. *Streptococcus zooepidemicus* is the bacteria most frequently isolated. *Rhodococcus equi*, commonly present in soil and the intestinal tract of horses, causes a particularly severe bacterial infection and may be associated with contagious outbreaks involving up to 50 percent of foals. Among *viruses* that cause neonatal pneumonia are those associated with rhinopneumonitis (EHV-1), equine influenza, and equine viral arteritis. Pneumonia also occurs secondarily as a complication of foal septicemia.

Factors that predispose a foal to pneumonia are failure to receive colostrum; foaling in overcrowded, cold, damp quarters; keeping a dirty stall by not removing soiled feed and bedding; and failure to keep mare vaccinations current. All of these can be avoided by good management.

Bacterial pneumonia usually starts insidiously with lack of appetite, inactivity and depression. Respiratory signs develop rapidly. They include high fever, diarrhea, rapid pulse, difficult breathing, watery eyes, cough and a thick nasal discharge.

Viral pneumonia is acquired in utero from an infected placenta. The foal is born weak and depressed and develops respiratory distress within the first 6 hours after birth. Most foals succumb to secondary bacterial infection and die within 7 days.

Treatment: Antibiotics are used to treat bacterial pneumonia and to prevent secondary infection in viral pneumonia. They should be started as soon as possible and changed subject to culture and sensitivity tests. *Rhodococcus equi* respond dramatically to the combined use of oral erythromycin and rifampin. Antibiotics should be continued throughout the course of the illness and then 1 week longer. Failure to complete a full course of antibiotics is the chief cause of relapse. Good nursing care, as described for foal septicemia, is essential. The foal should be kept in warm, dry quarters, treated for dehydration, and given anti-inflammatory drugs to reduce fever. Maintenance of hydration is critical in liquefying secretions and promoting their elimination.

The occurrence of viral pneumonia can be greatly reduced by keeping all vaccinations current. A recommended immunization schedule for broodmares and pregnant mares is shown in Table II, appendix.

TYZZER'S DISEASE

This rare bacterial infection of the liver occurs in foals 1 to 4 weeks of age. The causative organism is *Bacillus piliformis*. Infection appears to be acquired through the ingestion of contaminated soil and feed.

The onset is sudden, with high fever, diarrhea, collapse and seizures—all occurring within a matter of hours. In most cases the foal is found dead or in a coma.

Treatment is directed at supporting the foal with intravenous fluids, antibiotics and anticonvulsants. Unfortunately, the rapidity of the disease makes it difficult to initiate treatment in time to save the foal. The diagnosis is often made at autopsy.

NEUROLOGIC DISORDERS

NEONATAL MALADJUSTMENT SYNDROME

The neonatal maladjustment syndrome (NMS), also called the convulsive foal syndrome, barker, wanderer, or dummy foal syndrome, is believed to be caused by an episode of central nervous system *hypoxia* that occurs during or after delivery. The reason for the hypoxia is unknown in most cases. Typically the delivery is uneventful and the foal appears normal for the first hour or two of life. This period of normalcy distinguishes foals with NMS from foals born with brain trauma, since foals with brain injury are abnormal at birth.

Within hours, the foal with NMS loses his suckling reflex and begins to exhibit abnormal behavior. Signs of brain involvement include seizures with spasms of the head and extremities and thrashing of the legs. These convulsions are accompanied by pronounced respiratory distress. The foal emits a high-pitched whinnying or "barking" sound. The convulsive behavior either gets progressively worse or fluctuates from better to worse.

Another characteristic feature of NMS is loss of the righting reflex. The foal lies on his side and seems incapable of making the coordinated movements necessary to roll onto his sternum and stand.

At some point the foal usually passes into a dummy phase. While in this stage, he is inert and recumbent, does not respond to stimuli, appears blind, and loses affinity for his mother. Later the foal becomes extremely active and wanders about aimlessly walking into walls.

As a consequence of repeated siezures, some foals develop brain swelling as well as thermoregulatory, respiratory and circulatory failure. This leads to death, which occurs in about half the cases.

Foals suffering from actinobacillosis, foal septicemia, and neonatal meningitis often exhibit signs like those of NMS. Accordingly, blood cultures and laboratory studies should be run on all foals exhibiting dummy or wanderer-like behavior, since some of these foals will have an infectious basis.

Treatment: The first priority is to control seizures. This can often be accomplished simply by turning the foal onto his sternum. Helping him to stand may also be effective. Valium or an anti-epileptic drug should be given intravenously for seizures that don't respond to the above.

Treat hypothermia by covering the foal with a woolen or electric blanket. Provide a clean, well-padded surface for the foal to rest on. Remove all hay and straw. Nasal oxygen is of value for foals with respiratory distress.

Foals who did not receive colostrum should be treated for FPT. Good supportive care, as described in the treatment of foal septicemia, is critical for recovery. Gastric ulcers are common in convulsive foals. Prophylactic treatment with anti-ulcer medication may be advisable (see GASTRIC AND DUODENAL ULCERS earlier in this chapter).

Foals who survive the convulsive phase have a favorable outlook. Reflexes return gradually over 2 to 3 days, with the suckling reflex being the last to return. Complete recovery is likely if the foal has adequate IgG levels, does not develop foal septicemia, and is able to stand and suckle within 4 days of birth.

SHAKER FOAL SYNDROME

The shaker foal syndrome is a paralytic disease caused by the ingestion of spores of the bacteria *Clostridium botulinum* present in hay and feeds. The spores grow in the intestinal tract of the foal, releasing a powerful neurotoxin that causes the paralytic signs. The syndrome occurs exclusively in the Mid-Atlantic states and is most common in Kentucky, where soils contain a high concentration of spores.

Foals 1 to 8 weeks of age are most often affected. Symptoms appear 1 to 4 days after the ingestion of spores. The first sign is a generalized weakness. The foal walks with a stiff, stilted gait or may be found lying down. When forced to stand, the foal trembles and shakes before collapsing. Paralysis of the swallowing mechanism causes drooling and protrusion of the tongue. Liquids may dribble from the foal's nose and mouth. Death by respiratory paralysis or aspiration pneumonia occurs in 1 to 3 days. The mortality rate is 90 percent.

Treatment involves the administration of IV penicillin and equine type B botulinum antitoxin. Good supportive care is essential.

Prevention: Vaccination of pregnant mares with *C. botulinum* type B toxoid affords some protection against the shaker foal syndrome. Note that there are three types of neurotoxin. Vaccination protects against only type B (the most common) but does not protect against types A and C. Vaccination should be considered in endemic and high-risk areas, especially on farms where the shaker foal syndrome has occurred before. The immunization series is shown in Table III, appendix. In immunized broodmares, administer the annual booster 2 to 4 weeks before foaling. The product is not labeled for vaccination of foals.

TETANUS

Tetanus is caused by the bacteria *Clostridium tetani*. Horses are more susceptible to tetanus than other domestic animals. This is because horses possess less natural immunity and are subject to many tetanus-prone injuries.

Tetanus is a major concern for newborn and suckling foals. The bacteria enter at the umbilical stump or through an open wound. If conditions are favorable, the bacteria produce a potent neurotoxin that is transmitted along nerves and ascends to the spinal cord and produces a type of stiff paralysis. The neurotoxin is also absorbed locally and carried by the bloodstream to the brain.

Foals as young as 1 week of age can develop tetanus. The wound or site of infection can be difficult to locate. Early signs are vague stiffness in the front and rear legs, followed by spasms in the jaw, neck and hind limbs. The foal often adopts a "saw horse" stance, with his neck stretched out and head extended. The tail is held out stiffly like a pump handle, the ears are erect and the nostrils flared. In the final stage, the foal goes down, assumes an arched appearance, and dies of respiratory paralysis.

Treatment: Early surgical treatment of the infected wound or umbilical stump is critical in stopping the progress of the infection. This involves opening the area widely, removing all devitalized tissue, irrigating thoroughly, injecting penicillin into the wound, and leaving the wound open for adequate drainage. High doses of IV penicillin are routinely administered. Tetanus antitoxin, tranquilizers, muscle relaxants, intravenous fluids, and skilled nursing in a veterinary hospital alter the course for the better. If treatment is started early and the foal does not go down, the outlook for recovery is good.

Prevention: This is best accomplished by keeping the mare's immunizations current (see Table II, appendix). To ensure high levels of antibodies in the colostrum, a tetanus booster should be given 3 to 6 weeks before foaling. If the mare's vaccination history is unknown, or more than 5 years have elapsed since her last vaccination, the newborn foal should receive both tetanus antitoxin (1500 units) and tetanus toxoid at two different intramuscular injection sites. To complete the initial series, repeat the tetanus toxoid immunizations at 3 months and 4 months of age.

Tetanus antitoxin is no longer routinely administered to newborn foals because of the danger of causing serum hepatitis.

FAINTING FOAL SYNDROME (NARCOLEPSY-CATAPLEXY)

The fainting foal syndrome is a rare sleep disorder that occurs in newborn foals and is also seen in older horses. Signs are those of excessive daytime sleepiness accompanied by rapid eye movements as if the horse were dreaming. The head is held close to the ground, the eyes are closed, and occasionally you will hear snoring. A biochemical abnormality in the sleep-wake center of the brain stem is believed to be the cause. Although many breeds are affected, a familial incidence exists among Suffolk and Shetland ponies.

A typical fainting episode begins with the foal buckling at the knees and falling over on his side. Recovery occurs in minutes to hours. Between attacks, the foal appears normal. Shetland and Suffolk ponies may be affected for life. In most other breeds, symptoms rarely persist beyond a few weeks. Medication is available for treatment.

CEREBELLAR ABIOTROPHY

Incomplete neurological development of the cerebellum is an uncommon disease that occurs almost exclusively in horses of Arabian ancestry, although a similar disorder has been described in Gotland ponies and Oldenburg horses. The disease is caused by a decrease in the number and distribution of specialized neurotransmitter (Purkinje) cells in the part of the brain that controls balance and coordination.

Signs are occasionally present at birth, but usually appear in foals older than 6 months of age. Head tremors, incoordination, a staggering gait and a wide-based stance are characteristic. Often there is a peculiar type of gait called goosestepping, most pronounced in the front legs. The foal may buckle in the rear or fall over backward when made to back up.

There is no treatment. Mildly affected foals who acquire the disease as older individuals may learn to compensate and function normally.

OCCIPITO-ATLANTO-AXIAL MALFORMATION

Occipito-atlanto-axial malformation (OAAM) is a developmental abnormality involving the bone at the base of the skull and the first and second vertebrae in the neck. The effect of the malformation is to narrow the vertebral canal and compress the spinal cord. The disease occurs most often in Arabians and less frequently in Morgans and Standardbreds. It is inherited as an *autosomal* recessive trait. Both the sire and the dam must be carriers for the foal to be born with OAAM.

Affected foals are often born dead. In others, symptoms are present at birth or develop within the first few months of life. Signs include weakness; a jerky, uncoordinated gait; and paralysis of the legs. The neck is held in a characteristically stiff and erect position, like a "weather vane horse." Bending the neck to nurse, for example, can compress the cord and cause the foal to collapse.

Medical treatment is not effective. Surgical treatment involves removing bone over the top of the spinal cord to widen the vertebral canal. To be effective, surgical decompression must be performed before permanent damage occurs to the spinal cord.

CONGENITAL DEFECTS

Any disorder present at birth is referred to as congenital. Some congenital defects are genetically determined and others are acquired in utero. Still others are caused by accidents during labor and delivery. The following are among the most common congenital disorders in foals.

NEONATAL ISOERYTHROLYSIS

Neonatal isoerythrolysis (NI), also called hemolytic disease of the newborn, is of major concern because it affects 1 percent of Thoroughbred foals and may result in a fatal anemia. It begins shortly after the foal ingests colostrum-containing antibodies that destroy its red blood cells.

During pregnancy, fetal red cells (RBCs) can cross the placenta and enter the mother's circulation. When these cells are of a blood type that is incompatible with the mother's blood type, her immune system makes anti-RBC antibodies directed against these cells. These antibodies are present in colostrum and are ingested by the foal. The antibodies attack and destroy the foal's red cells, resulting in a hemolytic anemia.

The mare may have been exposed to fetal cells during a former pregnancy, or the exposure may have occurred during the current pregnancy. Receiving an incompatible blood transfusion is another cause of exposure.

Although there are more than 30 RBC types in the horse population, the majority of incompatibilities involve the factors Aa and Qa. In certain breeds, these types are virtually absent. For example, in the Standardbred, the Qa antigen does not exist. Thus Qa incompatibility does not occur in Standardbred horses.

Mares at low risk for producing foals with NI are those who possess Aa and Qa blood factors; because they possess these factors, the immune system will not make antibodies against them. Similarly, mares at risk are those who do not have these blood factors.

Anti-RBC antibodies in the mare's colostrum will not attack the foal's red cells if he possesses the same blood type as his mother. However, if he inherits his blood type from his sire, and if it is this type to which the mare was sensitized, then incompatibility exists and hemolytic anemia may develop.

Foals with isoerythrolysis are normal at birth. However, 24 to 36 hours after ingesting colostrum, some foals begin to experience signs of anemia, including lethargy and weakness. The severity of the anemia depends on a number of factors, including the blood types involved, the concentration of antibodies in the colostrum, and the amount of colostrum ingested. In many cases RBC destruction is slow, and few if any signs develop.

When RBC destruction is rapid and extensive, a sudden anemia develops. The heart and respiratory rates are greatly increased. The foal often lies with his chin resting on the ground while gasping for air. The breakdown of red cells releases free hemoglobin that turns the urine the color of dark tea. Jaundice develops as the liver converts the hemoglobin to bile. The bile then accumulates in the tissues and causes the mucous membranes of the gums and tongue to appear yellow. In a severe crisis, the foal may die before developing jaundice.

Treatment: The diagnosis is made by laboratory tests, including blood counts. By the time the condition is recognized, there is no need to stop the foal from nursing. The mare's milk no longer contains colostrum. In addition, the foal is no longer capable of absorbing maternal antibodies.

In the symptomatic foal, activity should be restricted. IV fluids are administered to maintain hydration and promote a large urine output. This helps to flush out hemoglobin before it can clog the kidneys. Antibiotics are often used to prevent secondary infection. Blood transfusions are required only if the anemia is life-threatening. The mare's own blood can be used, provided the serum is first siphoned off and the cells are washed. Otherwise, another horse of compatible blood type should be used as the donor. The prognosis is good if the anemia stabilizes and the foal does not develop secondary infection.

Prevention: If a mare has produced an NI foal in the past, you can test the mare's colostrum for anti-RBC antibodies that react with the foal's red cells. This should be done before allowing the foal to nurse. If the test is positive, give substitute colostrum by stomach tube and prohibit the foal from nursing for 24 hours. If the test is negative, the foal can nurse.

Some veterinarians find it simpler to omit the colostrum test, prohibit nursing for 24 hours, and treat all potential NI foals with substitute colostrum. If possible, the foal should stay with his mother. Nursing can be prevented by muzzling the foal or erecting a barrier that separates the foal from the mare.

PATENT URACHUS

This is a common congenital abnormality. The urachus is a tube in the umbilical cord that connects the foal's bladder to the mother's placenta. This channel closes at birth when the cord is severed.

The reopening of a closed urachus occurs in sick recumbent foals and in foals whose umbilical cord was cut or tied at birth. A foal with meconium colic who repeatedly strains to defecate is predisposed to reopening of the urachus.

When the urachus fails to close or reopens in the early neonatal period, urine dribbles from the umbilicus. The skin around the navel may become scalded and inflamed. This produces an excellent media for bacterial growth. Infection can ascend into the abdomen and produce peritonitis or foal septicemia; see Navel Ill above.

Treatment: Keep the umbilical area as clean as possible and apply a topical preparation such as Triple Antibiotic ointment™. Treat urine scalds with zinc oxide ointment. Oral antibiotics are routinely used to prevent ascending infection. The urachus can often be made to close by chemically cauterizing the opening and channel for about 2 inches in depth twice daily with silver nitrate sticks. If this is not successful within several days, surgical closure is indicated.

RUPTURED BLADDER

This occurs in somewhat less than 1 percent of newborns, primarily in colts. Some cases are caused by developmental weakness in the wall of the bladder along with bladder compression during delivery. An infected urachus that involves the bladder is another cause of bladder wall weakness. When the bladder ruptures, urine passes into the peritoneal cavity.

Signs begin within 36 hours of birth. The foal appears depressed, stops nursing, strains unsuccessfully to urinate or passes urine in small amounts. As the peritoneal cavity fills with urine, the abdomen becomes enlarged and pendulous. Ultimately the foal develops difficulty breathing, convulses and goes into coma.

Treatment: The diagnosis is made by *abdominocentesis* with the finding of urine in the peritoneal cavity. Treatment involves opening the abdomen and repairing the bladder. This should be done before the foal's condition has deteriorated. The foal often requires medical treatment in preparation for surgery.

LIMB DEFORMITIES

Angular limb deformities (knock-knees, bowlegs, bucked knees) and *flexural limb deformities* (contracted digital flexor tendons and clubfoot) are often of congenital origin. Less frequently they are acquired after birth as a result of improper exercise and nutrition. Some limb deformities are associated with Neonatal Hypothyroidism (see below).

These "windswept" legs should correct within the first week of life.

When a foal is born with weak or crooked legs, there is no need for immediate concern as long as the foal can stand and nurse. Most of these deviations correct spontaneously in the first few days after birth as muscle tone improves and ligaments strengthen. However, all cases of angular limb deformity that do not correct by 1 week of age should be seen by a veterinarian.

Foals who are normal at birth but develop angular limb deformities as they grow definitely require veterinary evaluation. Treatment may include hoof trimming, exercise restriction, and occasionally surgery. These foals should be evaluated before 6 weeks of age.

BROKEN RIBS

Ribs can be broken during passage through the birth canal and by efforts to administer CPR to a depressed foal. The signs are those of respiratory difficulty with a characteristic grunt heard as the foal breathes in.

Pneumothorax (collapse of the lung) will occur if the sharp edges of a broken rib puncture the lung. Pneumothorax usually is fatal.

If a newborn foal exhibits respiratory difficulty, notify your veterinarian.

COMBINED IMMUNODEFICIENCY DISEASE

Combined immunodeficiency disease (CID) is a fatal genetic disease of Arabian and part-Arabian foals characterized by a deficiency of B and T cell lymphocytes. These specialized cells manufacture IgM antibodies and are essential to the immune system. Affected foals are normal at birth. However, as maternally acquired immunity begins to wane, they become highly susceptible to infectious diseases, particularly adenovirus pneumonia. Once infected, they succumb rapidly. This usually occurs at 3 to 4 months of age.

The condition is inherited as an *autosomal* recessive trait. It is estimated that 25 percent of Arabian horses in the United States are carriers of the recessive gene and 3 percent of Arabian foals are born with the disease. Both the sire and dam must be carriers for the foal to be affected. Currently there is no way to identify carrier horses other than by testing their offspring.

The diagnosis is made by blood studies that reveal a lymphocyte count of less than 1000 per mL and an absence of serum immunoglobulin (IgM). Foals with the disease can be identified at birth if the blood sample is obtained before the foal nurses and acquires IgM antibodies from the dam.

Currently there is no treatment for CID. Early diagnosis can spare unnecessary medical expense and provide genetic counseling.

HEART DEFECTS

Although heart defects are not common, all forms of congenital heart disease (CHD) have been encountered in horses. More than one defect may exist at the same time. Mild defects may be asymptomatic and compatible with a normal life.

The most common abnormality is a ventricular septal defect, a hole in the wall that separates the two ventricles. Blood can flow from the right to the left side of the heart through this opening without going through the pulmonary circulation and receiving oxygen. Malformations of the heart valves also occur.

A patent ductus arteriosus is a persistent fetal artery joining the pulmonary artery to the aorta. This passageway is required in utero for blood to bypass the fetal lungs. The ductus closes shortly after birth. A coarse "machinery" murmur caused by a still patent ductus is normal up to the 7th day of life. If the ductus remains open, aortic blood shunting through the pulmonary circulation may produce severe pulmonary hypertension. The severity of the problem depends on the size of the shunt. Foals with a small shunt may live many years before developing exercise intolerance and heart failure. Horses with a large shunt develop pulmonary hypertension, followed by heart failure and death.

Signs of moderate to severe heart disease include poor growth; lethargy; exercise intolerance; rapid, heavy breathing; *cyanosis* and collapse.

There is no surgical treatment for CHD in horses. Most defects have a genetic basis. Horses with any type of CHD should be excluded from breeding programs.

NUTRITIONAL MYOPATHY (WHITE MUSCLE DISEASE)

Nutritional myopathy is a disease of foals from birth to 12 months of age. It is caused by a deficiency of selinium secondary to an inadequate intake of selenium by the mare during pregnancy and lactation. Selenium deficiency is particularly common on the East and West coasts as well as in the northern Midwest. One function of selenium is to enhance the absorption of vitamin E from the intestine. Accordingly, a vitamin E deficiency may also exist.

Signs of acute selenium deficiency include diarrhea, muscle pain and stiffness. In a severe case, the foal may develop respiratory insufficiency, followed by heart failure and death. A postmortem examination of muscle discloses abnormal fibers that appear white instead of red. Thus the disorder has been called white muscle disease.

Treatment: Injections containing selenium and vitamin E can be given by your veterinarian. If given during the first 2 weeks of life, the survival rate is about 50 percent.

Prevention: In areas where selenium deficiency is known to occur, trace-mineralized salt containing selenium should be fed free-choice as the only salt available. The mare should be given a selenium vitamin E injection during the last trimester of pregnancy, although this might not prevent a deficiency in all foals. It is also important to treat the foal with an injection shortly after birth. It might not be necessary to repeat the injection if the foal's creep and weanling feeds contain adequate amounts of selenium and vitamin E.

NEONATAL HYPOTHYROIDISM

Hypothyroidism in newborn foals is characterized by enlargement of the thyroid gland, along with signs described below. The disease, which begins in utero, usually is caused by feeding iodine supplements such as kelp seaweed to pregnant mares. Kelp contains high concentrations of iodine. Iodine interferes with production of thyroid hormone. Less frequently, hypothyroidism is caused by the mare grazing or consuming plants containing chemicals that block the activity of the thyroid. In either case, her foal does not manufacture or release enough thyroid hormone. Thyroid enlargement (goiter) is a compensatory effort to boost thyroid production by increasing the size of the gland.

Hypothyroid foals are weak and lethargic, suckle poorly, exhibit a lack of coordination and poor righting reflexes, and are often hypothermic. The enlarged thyroid in the neck may or may not be visible, depending on its size. Some newborns are asymptomatic but develop skeletal problems, including angular limb deformities and contracted tendons, at 2 weeks of age.

The diagnosis is made by thyroid function tests. Thyroid hormone replacement corrects the symptoms. However, limb deformities, if present, do not improve.

EYE DISEASES

Entropion is a birth defect in which the eyelids roll inward, causing the eyelashes to rub against the surface of the eye. *Ectropion* is the opposite condition; the eyelids are everted, exposing the eye to excessive drying. These two conditions should be treated to prevent permanent eye damage.

Congenital cataracts are present at birth and are the most common cause of blindness in foals.

Congenital night blindness is a hereditary condition found primarily in Appaloosas.

A foal can be born with one or both eyes smaller than normal. In some cases there is almost complete absence of an eye. A small eye is often associated with a congenital cataract.

CLEFT PALATE

The nasal and oral cavities are separated by the hard and soft palates. When these structures do not develop normally, there is a cleft in the roof of the mouth and an opening between the two cavities.

Foals with a cleft palate find it impossible to suckle. They cough, choke and gag, and are unable to swallow. Milk comes out the nose when the head is down. Surgical correction can be attempted. Results are best when the defect is small and is limited to the soft palate.

MALOCCLUSION PROBLEMS

Congenital malocclusions are caused by a disparity in the growth of the upper and lower jaws. Most malocclusions are apparent during the first weeks of life.

The bite or occlusion is determined by seeing how the upper and lower incisors meet. In a correct bite, the incisor teeth meet edge to edge. An incorrect bite is one in which the teeth meet in some other alignment.

A foal with a parrot mouth. Treatment should begin as soon as possible.

The most common jaw deformity is the *parrot mouth* or *overshot jaw*, in which the lower jaw is shorter than the upper jaw. In consequence, the upper incisors overhang the lowers. Because the upper incisors are unopposed, they grow long like rabbit teeth. When the malocclusion is restricted to the front teeth, it might not cause a problem. However, if the molar teeth are also out of alignment, they will not be ground down and will form hooks and sharp points. These hooks may interfere with chewing and injure the soft tissues of the mouth.

Parrot mouth can be treated by applying wire tension bands from the upper incisors to the first maxillary cheek teeth in an attempt to slow the rate of growth of the upper jaw. These braces can be left in place for several months and are removed when the bite is almost corrected. Results are best when treatment is started before 6 months of age.

The *sow mouth,* or *undershot jaw,* is the reverse of the parrot mouth. The upper jaw is shorter than the lower jaw so that the lower incisors project beyond the uppers like a bulldog. Sow mouth is less common than parrot mouth.

In horses, the upper dental arcade is always wider than the lower. In the *shear mouth,* this discrepancy is further exaggerated. This produces long, extremely sharp shearing edges on the cheek teeth.

Severe malocclusions can lead to mouth infections, poor mastication and impaired digestion of feed. This can compromise growth and development.

Foals with sharp edges on the cheek teeth will need to have the teeth rasped at frequent intervals. Feeding hard pellets or unprocessed grain may prolong the interval between treatments.

Malocclusions are genetically transmitted. Horses with such deformities should not be used for breeding.

COLON ABNORMALITIES

Absence of the anus *(anal atresia)* is a relatively common defect. The entire anus may fail to develop, or the rectum may extend down to a well-developed sphincter with a dimple at the site of the anal opening. These latter cases can be fixed with surgical correction.

Absence of part or all of the large colon occurs rarely. Surgical repair is difficult.

Aganglionosis is called lethal white foal syndrome because it occurs in white, blue-eyed, pink-skinned foals born to Paint Horses, the result of mating overo to overo. In this condition, the ganglion nerve plexus in the wall of the bowel does not develop. Lacking this plexus, the colon remains paralyzed and fails to contract. The foal develops a bowel obstruction in the first few days of life. Because the disease is irreversible, euthanasia is recommended to avoid unnecessary suffering.

APPENDIX

Table I **MINIMUM DAILY NUTRIENT REQUIREMENTS**

500 Kg (1,100 lb) Mature Weight*

Class of Horse	Weight kg	Lb	Energy Mcal	Protein gm	Lysine gm	Calcium gm	Phosphorus gm	Vitamin A (1,000 IU)	Potassium gm
Mature Horse, Maintenance	500	1,100	16.4	660	23	20	14	15	25
Mare, Late Pregnancy	500	1,100	18.8	866	30	37	28	30	31
Lactating, First 3 Months	500	1,100	28.3	1,400	50	56	36	30	46
Nursing Foal, 3 Months Age	155	341	14	650	28	35	25	7	12
Weanling, 6 Months Age	215	462	16	800	34	40	30	10	13
Yearling, 12 Months Age	325	715	20	900	38	40	30	15	18
2-Year-Old	450	990	18.8	800	32	24	18	20	23

*Source: *Nutrient Requirements of Horses*, 5th ed., National Research Council (1989).

For every 45 kg (100 lb) above and below 500 kg (1,100 lb), add or subtract 8 percent from the values given.

DE, mineral and vitamin requirements for light, medium and intense work can be estimated by increasing the maintenance requirements by 25, 50 and 100 percent, respectively.

Table II **IMMUNIZATION SCHEDULES**

(Minimum Recommended for All Horses)

Disease/vaccine	Foals/ Weanlings	Yearlings	Pleasure	Performance	Pregnant mares
Tetanus Toxoid*	3 mos, 4 mos	12 mos	Annually	Annually	Annually, 3–6 weeks before foaling
EEE, WEE	3 mos, 4 mos	12 mos, in spring	Annually, in spring	Annually, in spring	Annually, 3–6 weeks before foaling
Equine Influenza	3 mos, 4 mos, 5 mos, repeat every 3 mos	Every 3 mos	Biannually	Every 3 mos	Biannually, with booster 3–6weeks before foaling
Rhinopneumonitis (EHV-1 and EHV-4)	3 mos, 4 mos, repeat every 3 mos	Every 3 mos	Optional, biannually if elected	Every 3 mos	5th, 7th and 9th month of pregnancy (inactivated EHV-1 vaccine only)

*Administer after injury or surgery if the horse has not been vaccinated within the past 6 months.

All vaccines: Follow the recommendations of the manufacturer to avoid improper administration.

Table III ADDITIONAL IMMUNIZATIONS

(Recommended for Horses in Endemic and High Risk Areas)*

Disease/ Vaccine	Foals/ Weanlings	Yearlings	Pleasure	Performance	Broodmares
Rabies	A single vaccination at 12-16 wks	Annually	Annually	Annually	Annually before breeding; not while pregnant
Strangles	3 vaccinations at 3 wk intervals, beginning at 8 wks	Biannually	Biannually	Biannually	Biannually, 3-6 wks before foaling
Venezuelan (VEE)	3 mos, 4 mos	Annually, in spring	Annually, in spring	Annually, in spring	Annually, 3-6 wks before foaling
Potomac Fever	3 mos, 4 mos	Annually or biannually	Annually or biannually	Annually or biannually	Biannually, 3-6 wks before foaling
Equine Viral Arteritis	3 mos	Annually	Annually	Annually	3-6 weeks before breeding; not while pregnant
Botulism	Currently not recommended	Annually	Annually	Annually	3 vaccinations at 4 wk intervals, last dose 2-4 weeks before foaling

*To learn if the disease is endemic in your area, check with your veterinarian or state university extension agent.

All vaccines: Follow the recommendations of the manufacturer to avoid improper administration.

NORMAL PHYSIOLOGIC DATA

NORMAL RECTAL TEMPERATURE

Adult Horse (mares and stallions): 99.5° to 100°F (37.5° to 37.8°C)
Foal: 99 to 102°F (37.2° to 38.9°C)
Neonatal Foal (first hours of life): 99.5° to 100.5°F (37.5° to 38°C)

HOW TO TAKE YOUR HORSE'S TEMPERATURE

The only effective way to take your horse's temperature is by rectal thermometer. Large-animal bulb thermometers and digital thermometers are equally suitable. If using a bulb thermometer, shake it down until the bulb registers 96°F (35.5°C). Lubricate the bulb with petroleum jelly. Raise the horse's tail and hold it firmly; then gently insert the bulb into the anal canal with a twisting motion. Insert the thermometer 2 to 3 inches.

Hold the thermometer in place for 3 minutes. Remove the thermometer, wipe it clean, and read the temperature by the height of the silver column of mercury on the thermometer scale. If using a digital thermometer, follow the directions of the manufacturer.

Clean the thermometer with alcohol to prevent the transfer of disease.

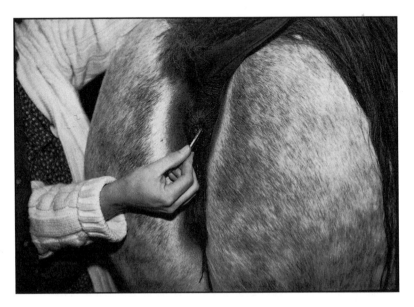

Lubricate the thermometer and insert it 2 to 3 inches into the anal canal. Wait 3 minutes and read.

PULSE

Normal: 44 (23–70) beats per minute
Nursing Foal (1 month of age): 70 to 90 beats per minute
Neonatal Foal (first hours of life): 80 to 130 beats per minute (average: 70)

HOW TO TAKE THE PULSE

The pulse, which is a reflection of the heart rate, can be taken at any point where a large artery is located just beneath the skin. The pulse rate is determined by counting the number of beats per minute. A convenient place to take the pulse is where the external maxillary artery crosses the lower border of the jaw. To locate the pulse, press lightly with the pads of your index and long fingers along the inside of the angle of the jawbone. The pulse is easiest to locate after the horse has been exercised.

The *digital* pulse is taken by palpating either branch of the digital artery, located on the inside or outside of the ankle just below the fetlock. Normally the digital pulse is barely detectable; however, it will be strong and bounding in acute laminitis.

Another way to count the pulse is to feel the beat of the heart itself. Place your hand on the left side of the horse's chest just above the point of the elbow. If the horse is not fat, you should be able to feel the impact of the heart with each contraction.

Taking the jaw pulse. With the first 2 fingers, feel along the inside of the angle of the jawbone.

It is difficult to take the pulse of a neonatal foal without using a stethoscope. If you do not have a stethoscope, place a palm on each side of the foal's chest beneath the elbows and feel for the heartbeat. If no heartbeat is felt, you can attempt to hear the heartbeat by drawing the left leg foreword and pressing your ear against the side of the foal's left chest behind his elbow.

RESPIRATORY RATE

Normal: 12 (10–30) breaths per minute
Neonatal Foal (first hours of life): 30 to 50 breaths per minute
Note: Determine the breathing rate by observing and counting the movements of the nostrils or flanks.

REPRODUCTIVE DATA

Puberty: 15 to 20 months
Natural Breeding Season (Northern Hemisphere): April to October
Operational Breeding Season (Northern Hemisphere): February 15 to July 15
Length of Estrous Cycle: 21 to 23 days
Length of Estrus (Heat): 4 to 8 days
Length of Diestrus: 14 days
Length of Gestation (Pregnancy): 327 to 357 days. (Average: 340 days)

DETERMINING YOUR HORSE'S WEIGHT

When weighing by scale is impractical, the weight of the horse can be estimated rather accurately by using a horse's weight tape, available from feed stores and tack shops. The tape is marked in pounds (or kilograms) of body weight as measured at the heart girth. In near-term pregnant mares, the tape underestimates weight by 150 to 200 pounds. This should be added to the result given by the tape.

If a commercial tape is not available, use an ordinary measuring tape and consult the information given in Table IV.

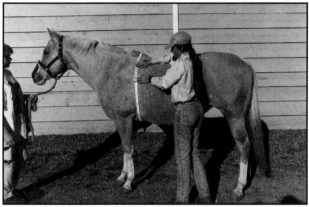

**Height-weight tapes are available at tack and feed stores.
Weight measurements are taken at the heart girth.**

Table IV   **DETERMINING YOUR HORSE'S WEIGHT**

| Heart Girth Length | | Weight | |
Inches	Centimeters	Pounds	Kilograms
30	76	100	45.5
40	102	200	91
45.5	116	300	136.5
50.5	128	400	182
55	140	500	227
58.5	148	600	273
61.5	156	700	318
64.5	164	800	364
67.5	171	900	409
70.5	178	1000	455
73	185	1100	500
75.5	192	1200	545
77.5	197	1300	591

BODY-CONDITION SCORE

Visual inspection is an effective way to estimate body condition. The following scoring system has been developed by equine nutritionists:

- **Poor** (Score 1): An extremely emaciated horse with prominent, bony ridges projecting beneath the skin in areas previously covered by muscle. No fatty tissue can be felt. If the condition is slightly better than extremely emaciated, give the horse a score of 2.

- **Thin to moderately thin** (Score 3 to 4): The horse has visible ribs, a slight ridge or prominence down its back over the spinous processes, and a prominent tail head. The withers, shoulders and neck are accentuated.

- **Well-conditioned** (Score 5): The back is level. The ribs can be felt easily but are not visible. There is a layer of fat around the tail head, and the withers are rounded over the spinous processes. The shoulders and neck blend smoothly into the body.

- **Moderately fleshy to fleshy** (Score 6 to 7): There is a discernible crease down the back. The ribs can still be felt, but there is a noticeable layer of spongy fat filling the spaces between them. The fat around the tail head is soft. Fat is apparent around the neck, over the withers, behind the shoulders and along the inner buttocks.

- **Obese to extremely obese** (Score 8 to 9): The above findings are more pronounced. The neck, in particular, is noticeably thickened. The ribs cannot be felt. Fat between the buttocks may rub together. The tail head bulges, the flanks are filled in flush, and the abdomen is potbellied.

Glossary

Abdominocentesis. The method of obtaining abdominal fluid through a large-bore needle or teat canula passed through the skin into the *peritoneal cavity*.

Abortion. Loss of the fetus between 45 and 300 days' *gestation*.

Abortion Storms. When a large number of mares simultaneously abort.

Accessory Sex Glands. The vesicular glands, prostate, and bulbourethral glands.

Anovulatory Follicles. *Follicles* that grow to a certain point but do not ovulate.

Anestrus. Lack of estrus or failure to cycle. The normal condition in winter.

Antibody. A protein substance produced by the immune system to neutralize the effects of an *antigen*.

Antigen. A substance recognized by the immune system that (if foreign to the body) may interfere with metabolism or cause a disease.

Artificial Insemination (AI). The process by which semen is collected from the male and inserted into the uterus of the female.

Artificial Vagina. A man-made receptacle used to stimulate ejaculation and receive the stallions's semen.

Assisted Breathing. Artificially filling and emptying the lungs to temporarily support the respiratory system.

Atrophy. The shrinkage in size of an organ or tissue due to disuse or death of cells.

Autosomal. All the chromosomes except the sex chromosomes.

Bactericidal. Capable of killing bacteria, as opposed to just inhibiting their growth.

Bean. A buildup of *smegma* in the urethral process of the *glans penis*.

Belling. Swelling of the head of the penis; normally occurs after *intromission*.

Benign. Refers to a growth that is not a cancer.

Bilateral. On both sides.

Biopsy. The removal of a piece of tissue for microscopic examination and diagnosis.

Bulbourethral Glands. *Accessory sex glands* in the male that produce seminal fluid.

Breeding Roll. A fabric-covered roll that is inserted between the mare and the stallion to limit the extent of penile *intromission*.

Canula. A blunt-nose hollow tube (medical instrument) used to drain a duct or body cavity.

Cardiac Massage. Compression of the heart, resulting in temporary support of the circulation.

Caslick's. The operation in which the lips of the mare's vagina are sutured together to prevent *pneumovagina* and vaginal contamination from manure.

Chorionic Villi. Finger-like projections ¹/₂ inch long found on the side of the placenta that makes contact with the *endometrium* of the pregnant uterus.

CL. *Corpus luteum.*

Colostrum. The first milk of the dam, containing the all-important maternal antibodies that protect the newborn foal for the first 3 months of life.

Conceptus. The products of conception resulting from the union between sperm and egg.

Conformation. How the various angles and parts of the body agree or harmonize with each other. A horse with good conformation has correct alignment and efficient movement in all gaits.

Congenital. Existing at birth, although not always clinically evident. Either genetically determined or acquired during or before delivery.

Corpus Cavernosum. The spongy erectile chambers of the shaft of the penis, which when filled with blood, result in erection.

Corpus Hemorrhagicum (CH). The contents of the *follicle* immediately after ovulation. The CH fills with blood and is transformed into the *corpus luteum.*

Corpus Luteum (CL). A growth that forms in the ovary at the site of ovulation. The CL manufactures progesterone, essential to the support of pregnancy.

CPR. Cardiopulmonary resuscitation; the combination of assisted breathing with cardiac massage.

Cryopreservation. The freezing and storing of sperm for later use.

Cyanosis. A blue color of the mucous membranes of the nose, lips and gums, caused by a deficiency of oxygen in the blood.

Cytology. The microscopic examination of cells to determine the cause of a disease.

Depression. Lack of interest to surroundings because of fever and illness; the equivalent of a severe headache.

Diestrus. Phase of the estrous cycle following ovulation during which the *corpus luteum* is present. Normally lasts 14 days. The period in which the mare rejects the stallion.

Ductus Deferens. The *vas deferens*, the tube that conveys sperm from the epididymis of the testicle up to the urethra.

Duodenum. The first 2 feet of small intestine just beyond the stomach.

Early Embryonic Death (EED). Loss of the embryo before 40 days' *gestation.*

eCG. Equine chorionic gonadotropin, the placental hormone that signals the *corpus luteum* to manufacture *progesterone.*

Edema. The accumulation of fluid beneath the skin along the midline of the abdomen; may also involve the *prepuce* and lower legs.

EIA. Equine infectious anemia.

Ejaculate. The total volume of *semen* emitted by the stallion during ejaculation.

Electrolytes. Sodium, chloride, potassium, bicarbonate, calcium, phosphorus and other minerals, the specific concentrations of which are required for cellular function.

ELISA. Enzyme-linked Immunosorbent Assay, a common serologic blood test used to detect *antibodies* to diseases.

Embryo. A *conceptus* younger than 40 days' *gestation*, before the stage of organ development.

Endemic. Dwelling in or native to a particular population.

Endometrial Cups. Specialized structures that develop in the placenta at 37 days' gestation. They produce *eCG*, essential to maintaining the pregnancy.

Endrometritis. Infection of the endometrium; the leading cause of mare infertility.

Endometrium. A layer of glandular tissue lining the cavity of the uterus.

Enteritis. Inflammation of the lining of the intestine, caused by a bacterial, parasitic or viral infection.

Epididymis. A coiled tube on top of the testicle that stores the sperm.

Estrous. The entire heat cycle, as determined from one ovulation to the next. It lasts an average of 21 to 23 days.

Estrus. The first phase of the estrous cycle, during which the mare is receptive to the stallion. It lasts an average of 4 to 7 days and ends with ovulation.

EVA. Equine viral arteritis.

Fertility. In stallions, the ability to impregnate the mare; in mares, the ability to conceive and carry a foal.

Fetotomy. The surgical division of a dead fetus into smaller parts to facilitate vaginal removal.

Fetus. A *conceptus* older than 40 days' gestation, generally after the stage of organ development.

Fibrosis. The replacement of glandular tissue by scar tissue.

Flagging. The up and down pumping action of the stallion's tail that occurs with ejaculation.

Flehmen Response. A curling of the stallion's upper lip in response to the presence of a mare in heat.

Follicle. A growth within the ovary that contains an egg.

Fresh Semen. Semen inseminated within 12 hours of collection.

FSH. Follicle-stimulating hormone, produced by the pituitary gland. It causes the ovaries to produce egg *follicles*.

Gastroscopy. The procedure during which the interior of the esophagus and stomach is viewed through a fiberoptic instrument passed through the horse's nose.

Gelding. Neutering a male by removing his testicles.

Gestation. The length of pregnancy. The period from conception to birth, lasting an average of 340 days, with a range of 327 to 357 days.

Glans Penis. The head of the penis; it contains erectile tissue that is separate from the shaft of the penis.

Goiter. An enlarged thyroid gland; associated with hypothyroidism.

Gonadotropins. Hormones released from the pituitary gland or placenta, acting on the ovaries or testicles to cause them to manufacture and release the sex hormones.

HCG. Human chorionic gonadotropin, a hormone given to induce *ovulation* in mares or stimulate the secretion of *testosterone* from the testicles of males.

Hematoma. A collection of clotted blood beneath the skin or at the site of injury or surgical wound.

Hermaphrodite. An individual having both testes and ovaries (rare). See the more common *pseudohermaphrodite*.

Histology. The microscopic study of the structure of tissue to determine the cause of disease.

Hydrotherapy. Cold water delivered to the site of injury using a shower head or nozzle.

Hypertrophy. Enlargement of an organ or tissue; an increase in size and volume.

Hypothalmus. A hormone center in the brain that controls the secretions of the pituitary gland.

Hypoxia. Lack of oxygen in the blood. If untreated, it results in coma and death.

IM. Intramuscular. An injection given into the muscle.

In Vitro Fertilization. A procedure in which an unfertilized egg is removed from the mare's uterus and fertilized with stallion's sperm in a glass dish.

Incarceration. The trapping of an organ or part of an organ within a closed space.

Incompetent Cervix. A cervix that does not close tightly during diestrus or pregnancy.

Infertility. Absence of *fertility*.

Intromission. The insertion of the stallion's penis into the mare's vagina during the act of breeding.

Involution. The shrinkage of the uterus back to normal size after delivery of the foal.

IV. Intravenous. An injection given into a vein.

Karyotype. The appearance (including number, size and shape) of the set of paired chromosomes possessed by a specific horse.

Laminitis. Inflammation of the inner structures of the foot. Same as founder.

Lesion. Refers to a change in the appearance of tissue caused by an injury or specific disease.

LH. Luteinizing hormone, produced by the pituitary. It causes an ovarian *follicle* to mature and ovulate.

Libido. Interest in the opposite sex; same as sex drive.

Live Cover. The act of mating between a mare and a stallion.

Luteal Activity. Refers to the influence of the *corpus luteum*, particularly to the effects of *progesterone* produced by CL.

Luteal Phase. The phase of the estrous cycle during which the *corpus luteum* is present in the ovary and manufacturing *progesterone*. Same as *diestrus*.

Luteolysis. The process that results in the regression and disappearance of the *corpus luteum*. Accompanied by a fall in serum *progesterone*.

Malignant Tumor. One that is harmful or dangerous and that can cause death. It can spread, or *metastasize* throughout the body.

Malignant. Refers to a growth that is a cancer.

Meconium. A pasty green material that makes up the first stools of the newborn.

Metastasize. The spreading of a cancer from its site of origin to other parts of the body.

Mucopurulent. Secretions that contain mucus and pus, usually indicative of bacterial infection.

Multiparous. Multiple pregnancies; used in reference to having been pregnant more than once.

Neoplasm. Any growth on the surface of the body or within an organ. Includes noncancerous and cancerous growths.

NSAID. Nonsteroidal anti-inflammatory drugs, such as phenylbutazone (Butazolidin™) and flunixin meglumine (Banamine™).

Ovaries. The paired organs that produce the female sex cells (eggs).

Oviduct. The tube that carries the egg from the ovary to the uterine horn.

Ovulation. The process during which the egg follicle releases the egg into the *ovulation fossa* and *oviduct*.

Ovulation Fossa. A recess in the lower pole of the ovary beneath the capsule into which the egg is deposited during *ovulation*.

Parturition. Giving birth; the period covered by labor and delivery.

Pathogenic. Having the potential to cause disease.

Pathogens. Bacteria capable of causing disease in a susceptible host.

Pelvic Brim. An important bony landmark in the mare's perineum. It can be felt below the floor of the vagina by finger palpation.

Perineal Body. A supporting group of muscles between the anus and vagina.

Perineum. The area between the anus and bottom of the vulva.

Peritoneal Cavity. The abdominal cavity, containing organs of the intestinal, urinary and reproductive tracts.

Placentitis. Infection of the placenta, usually caused by bacteria that ascend into the uterus through the cervix. Results in abortion or the birth of an infected foal.

PMNs. Polymorphonuclear leukocytes, inflammatory cells that make up pus.

Pneumothorax. Air in the chest caused by a tear in the lung or a wound in the chest wall. The lung collapses, resulting in respiratory distress.

Pneumovagina. An acquired state of anatomy in the *perineum* that allows air and fecal contaminants to enter the vagina and infect the uterus.

Postpartm. The 30-day period that follows foaling.

Potency. The *fertility* of the stallion as determined by the total number of progressively motile sperm he produces per *ejaculate*.

Premature Foal. A foal born alive after 300 days' but before 320 days' gestation.

Prepuce. The sheath of skin that surrounds the glans or head of the penis.

Priapism. A condition characterised by persistent erection without sexual arousal.

Primiparous. Describes a mare in her first pregnancy.

Progesterone. The pregnancy hormone. It is also responsible for the *diestrus* behavior of the mare.

Prolonged Diestrus. A *diestrus* phase of the *estrous* cycle that lasts more than 14 days. Also called prolonged luteal phase.

Prostaglandin. A hormone found in the lining of the uterus. Also refers to drugs containing the hormone.

Prostate. An accessory sex gland in the male that produces seminal fluid.

Pseudohermaphrodite. An individual whose external genitals resemble one sex, while the gonads (testes or ovaries) are those of the opposite sex.

Pustules. Small raised bumps or blisters containing pus, found on the skin or mucous membranes.

Reflux. A reversal in the normal direction of flow.

Scalds. Areas of inflamed or ulcerated skin caused by repeated or continuous contact with urine or feces.

Scrotum. The bag of skin and connective tissue that surrounds and supports the testicles.

Semen. The contents of the *ejaculate*, containing sperm cells, gel, and the secretions of the *accessory sex glands*.

Septicemia. (Infection throughout the system.) Refers to the invasion of the bloodstream and internal organs by bacteria, viruses and other pathogens.

Septicemic. Refers to the stage of infection in which the organism is found in the bloodstream.

Seropositive and Seronegative. Refers to the results of blood tests taken to determine if antibodies to disease-producing bacteria or viruses are present in the blood. A seropositive test result indicates that a horse has been exposed and has developed immunity.

Smegma. A thick, waxy substance secreted by sebaceous glands and flaking skin.

Spermatogenesis. The production of sperm by the testicles.

Sporadic. Isolated, occasional or infrequent.

SQ. Subcutaneous. An injection given subcutaneously, or beneath the skin.

Standing Heat. *Estrus.* The state in which the mare is ready and willing to breed, and stands fast for the stallion.

Stillbirth. A foal born dead after 300 days' *gestation*.

Strongyles. A species of roundworm that are among the most harmful and damaging intestinal parasites of horses.

Subfertility. Less than normal *fertility*.

Superinfection. A second infection on top of (or following) the first infection.

Systemic. System-wide. Often used in reference to the giving of a drug by the oral, intramuscular or intravenous route, as opposed to its local or topical application.

Teats. The two nipples of the mare's udder.

Testes. The testicles, or male gonads.

Teratogenic. That which causes developmental malformations in the fetus.

Testosterone. The male hormone, produced by the testicles; it is responsible for the secondary sex characteristics of the stallion.

Tone. Refers to the degree of firmness or softness of the uterus and cervix.

Torsion. The twisting of an organ and its blood supply, resulting in insufficient blood flow and death of that organ.

Toxemia. A condition caused by the spreading of a toxin (or the byproducts of inflammation) throughout the body.

Transitional Heat Periods. Estrous cycles that occur at the beginning and end of the breeding season; often associated with *anovulatory follicles*.

TSW. Total scrotal width, a measurement of the size of the stallion's testicles.

Tumor. Any growth or swelling. A true growth is called a *neoplasm*.

Type. A set of distinguishing physical characteristics that gives a purebred horse the defined look of its breed.

Ultrasonography. A diagnostic procedure that uses high-frequency sound waves to map a picture of an organ or structure beneath the surface of the body. Same as an ultrasound exam.

Unilateral. On one side only.

Urachus. A tube that connects the foal's bladder to the mother's placenta while the foal is in the uterus.

Urethral Diverticulum. A pocket in the *glans penis* above the urethra that contains an accumulation of *smegma*.

Urethral Process. The opening of the urethra at the end of the penis; it protrudes slightly beyond the head during erection.

Urovagina. The result of urine *refluxing* into the vagina and pooling around the cervix. Causes vaginal and uterine infection.

Uterine Inertia. Failure of the uterus to contract during foaling.

Vaginal Fornix. The deepest part of the vagina.

Vaginal Vestibule. The outer one-third of the vagina, extending from the labia to the beginning of the vaginal canal.

Vas Deferens. Same as *ductus deferens*.

Ventral Edema. The accumulation of fluid beneath the skin along the midline of the abdomen; may also involve the *prepuce* and lower legs.

Vesicle. A fluid-filled sac containing a young embryo.

Vulva. The labia (lips) of the vagina.

Vulvar Cleft. The split formed by the meeting of the vulvar lips. In the adult mare, the vulvar cleft averages 12 cm in length.

Vulvovaginal Fold. A landmark in the vaginal canal at the site of the hymen. In broodmares, it is marked by a flap of wrinkled tissue that runs horizontally across the vaginal floor and covers the opening of the urethra.

Vulvovaginal Sphincter. A mechanism that uses the vulvovaginal fold, reinforced by vaginal constrictor muscles, to close the vaginal canal.

Waxing. The formation of a honey-colored bead of colostrum at the end of each teat; usually indicates that foaling will occur within 24 hours.

Windsucking. The process in which air is actively sucked into the vagina as a consequence of *pneumovagina*.

Winking. The presentation of the clitoris that occurs in *standing heat*.

INDEX

NOTE: *Pages shown in boldface contain detailed coverage of the item.*